# Looking Down the Tree

# Looking Down the Tree

## The Evolutionary Biology of Human Origins

MITCHELL B. CRUZAN

*Portland State University*
*Portland, OR*
*United States of America (the)*

OXFORD
UNIVERSITY PRESS

# OXFORD
## UNIVERSITY PRESS

Oxford University Press is a department of the University of Oxford.
It furthers the University's objective of excellence in research, scholarship,
and education by publishing worldwide. Oxford is a registered trademark of
Oxford University Press in the UK and in certain other countries.

Published in the United States of America by Oxford University Press
198 Madison Avenue, New York, NY 10016, United States of America.





Library of Congress Cataloging-in-Publication Data
Names: Cruzan, Mitchell B. author
Title: Looking down the tree : the evolutionary Biology of human origins /
Mitchell B. Cruzan, Portland State University, Portland,
OR, United States of America.
Other titles: Evolutionary Biology of human origins
Description: New York, NY : Oxford University Press, [2025] |
Includes bibliographical references and index.
Identifiers: LCCN 2025015911 (print) | LCCN 2025015912 (ebook) |
ISBN 9780197805169 paperback | ISBN 9780197805152 hardback |
ISBN 9780197805183 epub | ISBN 9780197805190
Subjects: LCSH: Evolution (Biology) | Human evolution
Classification: LCC QH366.2 .C745 2025 (print) |
LCC QH366.2 (ebook) |
DDC 576.8—dc23/eng/20250728
LC record available at https://lccn.loc.gov/2025015911
LC ebook record available at https://lccn.loc.gov/2025015912

DOI: 10.1093/9780197805190.001.0001


Paperback printed by Marquis Book Printing, Canada

Hardback printed by Lightning Source, Inc., United States of America

The manufacturer's authorized representative in the EU for product safety is
Oxford University Press España S.A., Parque Empresarial San Fernando de Henares,
Avenida de Castilla, 2 – 28830 Madrid (www.oup.es/en).

*For all my students, past, present, and future—I will never stop learning from you.*

# Contents

# Preface

Why do humans look the way we do? It seems like we know a lot more about why other animals look different from each other. From numerous studies we have a good understanding of many animal characteristics—the giraffe's long neck, the digging claws of animals that burrow underground, stripes on zebras, the bright colors or camouflage of many insects and birds, the colors and patterns on flowers, and many others. But we seem to have less of an understanding of many human characteristics. Why are we bipedal when most other animals walk on four legs? Why don't we have fur all over our bodies like our closest relatives, chimpanzees, bonobos, and gorillas?

A number of books have discussed the origin of human characteristics and behaviors that are not evident from fossils, and most of these have focused on human sexual characteristics. The most well known include early books such as *The Naked Ape* by Desmond Morris and *The Third Chimpanzee* and *Why Is Sex Fun?* by Jared Diamond. More recent works include Alan Dixson's *Sexual Selection and the Origins of Human Mating Systems*, Peter Gray and Justin Garcia's *Evolution and Human Sexual Behavior*, Matt Ridley's *The Red Queen*, Robert Martin's *How We Do It*, and Robin Baker's *Sperm Wars*. These books discuss some of the same questions I address concerning human sexuality and the origin of our unique traits; however, much has changed in recent decades, and we now have access to a large amount of information—especially in the area of human genomics—that was not previously considered. In this book I provide novel insights into the origin of our unique characteristics, appearance, and behaviors, including the origin of human mating systems (i.e., monogamy), bipedalism, the lack of fur over most of our bodies, bulbous breasts in women, female orgasm, penile morphology, sperm competition, homosexuality, sexual and gender identity, altruism and cooperation, and much more.

While some other books purport to discuss the evolution of human traits and behaviors, all of them draw heavily on studies conducted in modern societies. We know that characteristics unique to any species are usually adaptations to their native environment, so it's confusing that other authors have not considered the conditions faced by our ancestors. It's harder to understand our own evolution when we only consider our modern environments, which are very different than the habitats where our unique characteristics originated—the dry forests and savannas of eastern Africa. Behavioral ecologists understand that the behaviors of captive animals are not the same as animals' behaviors in the wild.

In many respects, modern humans living in industrialized nations are captives in artificial environments they have created for themselves. This is not universally true, as many societies around the world do not enjoy the comforts of modern structures and technologies and, consequently, expected lifespan is much shorter. People living under these *more natural* conditions seem well adapted to their environments, but the same is not true for people who live with technologies that buffer them from the forces of nature; *their behaviors and physical characteristics seem out of place. The form of their bodies and the features that define all of us as humans developed over millions of years, but without a natural context, it is difficult to understand how selection led to our modern appearance and behaviors.*

In this book we explore aspects of human evolution that are not usually discussed in textbooks or biology classes. We know much about our history from bones and DNA, but these studies do not really tell us about the appearance of our ancestors and the origin of characteristics that are not preserved in the fossil record—the fleshy parts and behaviors of our ancestors. We can gain insights into our evolution by understanding the pressures of natural selection that our ancestors experienced—by viewing our species as an evolutionary biologist would. Evolutionary biologists are interested in the processes of evolution—the forces that shaped our ancestors and led to the unique appearance and behaviors that we have inherited.

It's difficult to imagine how challenging life was for our early ancestors. The hot and dry savannas of eastern Africa were harsh, and death from predators and disease was common. To understand how our appearance came about, we need to evaluate how each characteristic helped improve the fitness of our ancestors; their survival, and reproduction. Everything that makes us different—from bipedalism and loss of body hair to the enlarged breasts and buttocks of women, our social structures and behaviors, and the transmission of knowledge and skills through cultural evolution—contributed to improved survival and reproduction and the spread of our ancestors across the globe. In this book we analyze the characteristics and behaviors of humans in their natural habitat, just as we would any other species. By applying the principles of evolutionary biology, we develop a framework through which to understand our history and the origins of our unique appearance and behaviors.

To facilitate an understanding of the origin of human traits, I draw upon evidence and inferences from fossils, genomics, phylogenetics, coalescence theory, analyses of calcium isotopes, and the anatomy and physiology of our ancestors and other animals. I interpret these observations in the contexts of comparative biology, natural and sexual selection, evolutionary constraints, inbreeding and inclusive fitness, and genetic and cultural evolution. The story of our past that we piece together provides a novel view of how savanna habitats favored a unique set of adaptations including bipedalism and the loss of fur in our early australop-ithecine ancestors. Other characteristics such as bulbous breasts, female orgasm, monogamy, large penis size, cooperation, and exclusive homosexuality were out-comes of our infants becoming increasingly underdeveloped at birth (altricial) as head size increased and cooperation was favored by inbreeding and inclusive fitness in the clans of our *Homo* ancestors.[1] We end this discussion in the last chapter by examining how different traits interact and sometimes have cascad-ing effects as one trait leads to another. We evaluate the role of cultural evolution, which led to selection for larger brain volume, as the transmission of skills and knowledge became increasingly important. The results of this exercise provide a comprehensive view of the origin of our unique appearance and behaviors.

Not all readers will appreciate the approach I have taken to understand the origins of our species and, generally, the origins of life on Earth. One of the most outstanding characteristics of humans is our immense capacity for imagi-nation. We often find fanciful and implausible explanations of phenomena and events much more attractive than those offered by science. Our leanings toward the phantastic are buoyed by an expansive diversity of science fiction literature, movies, and series; some of which have become intrinsic to the cultures of many of our societies (e.g., *Star Trek* and *Star Wars*). It is often more exciting to imag-ine a fictional origin of our species than to consider the hard facts and logical explanations that science has to offer. My suggestion for readers is to retain your imaginary and fantastic views of the origins of life across the universe while appreciating the account I offer as one that is grounded in factual information and the logical arguments of science.

Dry descriptions of facts and analyses of characteristics can create a detached view that falls short of the deep appreciation of the challenges that our ancestors faced. Since it's sometimes easier to build understanding through storytelling, I've included the story of a fictional character who lived perhaps 70,000 years ago. The episodes of her life help illustrate how the characteristics and behav-iors of early humans contributed to their survival. At the beginning of each chapter, vignettes set the context for our exploration of why we look and act the way we do.

---

[1] Species in the same genus, *Homo*, that pre-dated the origin of *Homo sapiens* (abbreviated as *H. sapiens*), for example, *Homo neanderthalensis* for Neanderthals.

# Acknowledgments

I am indebted to a number of individuals who have provided valuable feedback, suggestions, and edits including Glenn Branch, Richard Kliman, Daniel Krumm, Anna Mele, Luis Ruedas, Rachel (Bice) Williams, and Desmond Willson.

# Prologue

*There once was a person who lived some 70,000 years ago—we can call her Launua. She worked daily along with her partner to provide food and shelter for her family and other members of her tight-knit community—her clan. Together, they wandered across the savannas of eastern Africa and took shelter wherever they could find it. It was a tough life, but Launua could rely on everyone in her clan because they were all part of the same family; if they were not a parent, brother, or sister, they were certainly an aunt, an uncle, or a cousin. She lived in a place that held many dangers from predators and disease. Her clan had no medicine or knowledge of healing, so even minor injuries could result in infection and death. For women, life was especially difficult, because many died from complications during childbirth. Death was so common that most people did not live past the age of forty. But there was little time for mourning; their survival depended on the daily struggle to find enough food from the roots and grains they gathered and from the occasional animal they were able to hunt and kill. When local sources of edible plants and game became scarce, they were forced to move. They carried their shelters and the tools they had fashioned from wood, bone, and stone. Their progress was slow because they had to care for their young along the way. They searched the arid savanna landscape for a suitable place to live; one that was near a reliable source of water and offered good foraging and hunting. As they traveled through the dry woodlands and grasslands, they felt exposed and were constantly on the lookout for predators.*

# 1
# How we do science

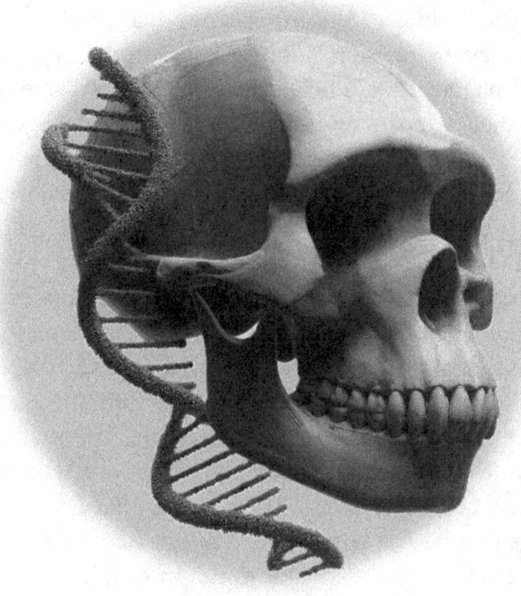

To start developing an understanding of human evolution, we first need to talk about the science that we rely on to discover the biology, ecology, and origins of all plants and animals. With science we can ask questions (*hypotheses*) and develop constructs of understanding (*scientific theories*) that guide our conclusions based on evidence.[1] One of the most significant inventions of modern societies is the *scientific method*,[2] which is the best tool we have to obtain the truth. Our legal and judicial systems can't compare. Unlike lawyers, scientists can't cherry-pick evidence to support their arguments—they have to consider all the evidence before they can draw any conclusions. Even the decisions made by the United States Supreme Court and decisions by high courts in other countries are often tainted by ideologies and individual bias because there is no

[1] Factual information from the physical world.
[2] "Scientific Method," Oxford Dictionaries: British and World English, 2016.

*Looking Down the Tree*. Mitchell B. Cruzan, Oxford University Press. © Oxford University Press (2025).
DOI: 10.1093/9780197805190.003.0001

requirement for justices to address the entirety of previous legal decisions—they can pick and choose as they wish.

Science isn't perfect, but it strives to be. Scientists are often called upon to review their colleagues' work, including manuscripts submitted to scientific journals for publication and grant proposals submitted to private and federal agencies for funding. The reviewers' comments and criticisms are then considered as part of the decision process for publication or funding. Usually this results in requests for revisions; the authors have to address all of these comments in their resubmittal of the proposal or manuscript, and it may take several rounds of review/revision before the manuscript or proposal is considered acceptable. Imagine if our judicial system worked the same way—how many faulty decisions that stripped away the rights of individuals would have been avoided? In science, failing to cite previous work—to take the whole of known evidence into account before making conclusions—can lead to the rejection of a manuscript or proposal. This system of peer review is the "gold standard" that scientists adhere to—something that science deniers fail to recognize. But even with such rigorous review, occasionally faulty work and biased interpretations are published. The scientific process is not perfect, but it's pretty much as close as we can get to the best system to find the truth.

Science is not dogmatic. There is no indelible scripture that is followed in the gathering and interpretation of *data*—the factual information that fuels the scientific process. Even Darwin's book *On the Origin of Species* is not taken as dogma. As we will see, there were a lot of things that Darwin did not know, and some things that he just got wrong. There is no "bible" of evolution that imposes dogmatic views on the natural world. In fact, it's exactly the opposite; the observations and data we collect inform the script of our understanding of life on Earth. If there were such a thing as the "book of science," it would be a dynamic document, not a static one. Imagine that new discoveries—reported in a single study—were written in this imaginary book in light pencil. Later, contradictory information might be published, in which case the original words would be erased. But if additional studies confirmed the original conclusions, then the words written in light pencil would become darker, and perhaps would be elaborated. As more evidence is gathered, the original ideas become more substantial, and with large amounts of evidence, they may eventually become indelible. Throughout this process, the original ideas are expanded and elaborated as our understanding of the natural world improves; with each new experiment we are able to ask more sophisticated and detailed questions, which leads to more experiments as further elaborations of our understanding become part of the book of science.

Scientists ask questions and conduct experiments within frameworks of widely accepted theories. But what is a theory? Most people use the word "theory" to describe something that is uncertain, but a *scientific theory* is a system of understanding that is supported by large numbers of experiments and large amounts of data.[3] When most people say they have a "theory" about something, they really mean a *hypothesis*. It's fine to use the word "theory" as a guess or idea in everyday discussions, but when we are talking about science, we need to say "hypothesis" instead. In science, we reserve "theory" for frameworks of understanding how nature works that are supported by large amounts of evidence.

Many people are skeptical about science when it does not agree with their worldview. It seems like challenges to conclusions made by scientists have become more common in recent decades, but science denial is nothing new. For example, the arguments made by creationists (including proponents of *intelligent design*)—who believed that life on Earth was too complex to have come about by natural processes—stem from the design arguments first made by William Paley, a natural theologian of the 1800s. Paley presented organs such as the eye and the workings of the solar system as too complex to have come about without the influence of a Creator. His arguments even convinced a young Charles Darwin during his early years as a student at Christ's College, Cambridge, where Darwin was studying to become an Anglican parson. This was before his famous voyage and, later, the development of his theory, which dramatically changed our understanding of biology and the diversity of life on Earth. Modern proponents of intelligent design have applied Paley's logic to the complexity of metabolic pathways within cells,[4] but to make those arguments, they have to ignore large amounts of evidence to the contrary. Consequently, intelligent design is not accepted as science and cannot be applied to improve our understanding of nature or the development of modern medicine, biotechnology, or agriculture. Darwin's principles of natural selection and evolution, on the other hand, have been applied across many disciplines outside of the biological sciences,[5] including computer science, software engineering, and the development of artificial intelligence algorithms.[6]

We say that scientists are supposed to be objective and consider all the evidence, but the opinions of scientists can sometimes be influenced by cultural or personal biases. Even Darwin was influenced by cultural bias when he wrote

---

[3] Note that this excludes ideas that are proposed as "theories" but fail to be supported by substantial evidence.

[4] M. J. Behe, *Darwin's Black Box: The Biochemical Challenge to Evolution*, 2nd ed. (Free Press, 2006).

[5] N. A. Johnson, *Darwin's Reach: 21st Century Applications of Evolutionary Biology* (Routledge, 2022).

[6] R. Miikkulainen, "Creative AI through Evolutionary Computation: Principles and Examples," *SN Computer Science* 2 (2021): 163.

that women and people of color had inferior intelligence.[7] It's important to recognize that scientists are not immune to the biases of the societies they live in. Darwin and other scholars of his day should have been more careful, but they must have thought they were saying things that were based on evidence. At the time, there were no comprehensive studies testing intelligence, so the opinion they expressed was based on casual observations of behavior. They did not consider the fact that people of various gender identities and from different cultural backgrounds may not have had the same educational opportunities. They didn't bother to examine their own biases, and of course they should have. But maybe we should be careful about blaming individuals for the biases of their entire culture.

Keep in mind that people used to believe all sorts of crazy things, and some still do. Not that many years before Darwin it was widely accepted that creatures such as mermaids and sea dragons were real. Carl Linnaeus even included descriptions and the scientific names of unicorns and sea dragons, along with a number of other mythical creatures, in his list of all known *species*.[8] It was common to believe such things in the 1700s, and the same could be happening today; scientists are not immune to social biases. It is possible that people in the future will look back on some of the things we believe today and consider them to be just as naïve as beliefs that were held 300 years ago.

It's important to realize that all people—even scientists—have biases based on their life experiences. This even affects the way scientists view each other's research. It's a well-known fact that it is more difficult to publish results that don't match the expectations of the majority, or how scientists think that nature works. To understand this better, let me tell you a story—a true story—about Barbara McClintock and her research on corn genetics.

In the mid-twentieth century, McClintock found that the genetic color characteristics of corn kernels she was following from one generation to the next seemed to "jump" from one chromosome to another. She knew this because of the careful breeding she had done, the replication of the crosses, and the large amount of data she had collected on this unique phenomenon. But she anticipated that her results would be met with skepticism because they did not follow expectations from the model of inheritance that Gregor Mendel had described late in the nineteenth century; a model that had been validated so many times that it was just assumed to always be true.

At first, McClintock's results on jumping genes were not believed at all— even to the point where reportedly people ridiculed her work at the conferences

---

[7] P. Murphy, "Reevaluating Female 'Inferiority': Sarah Grand versus Charles Darwin," *Victorian Literature and Culture* 26 (1998): 221–36.

[8] A group of morphologically and genetically distinct individuals capable of interbreeding with each other but generally not with members of other species.

she attended. Her lecture at the prestigious Cold Spring Harbor symposium was met with dead silence, and she described the reactions of her colleagues as "puzzlement, and even hostility." It did not help that she was a somewhat soft-spoken woman working in a field that was vastly dominated by men. But she persisted in conducting her experiments. Eventually more of her colleagues became convinced, but they said that maybe this sort of thing only happens in corn.

Although her work was eventually published,[9] it was largely ignored until the same phenomenon was observed in bacteria[10]—some twenty years after McClintock's publications on jumping genes. But instead of being called jumping genes, they were named *transposable elements.* It was a triumph for her that the value of her discovery was finally recognized. We now know that these small genetic fragments, which can move around within and among genomes, are extremely common. Since their prevalence was confirmed in the 1970s, there have been numerous studies on the molecular genetics and evolution of transposable elements, and they have been harnessed as a major tool for genetic manipulation and experimentation. In recognition of her efforts, McClintock was awarded the Nobel Prize in Physiology or Medicine in 1983—some thirty years after her original work on jumping genes was published.[11]

Why did McClintock persist in conducting experiment after experiment for so many years, even though she knew the scientific community was reluctant to accept her results? Like any good scientist, McClintock was driven by her desire for knowledge. She knew that the phenomenon she was observing was real because the data told her so. *There is tangible truth in data.* Data are evidence based on observations in the physical world that represent the currency that drives science. Data are the foundation of the scientific method; they support the process of inquiry, provide a means to test hypotheses, and support the conclusions drawn. These nuggets of truth are irrefutable.

McClintock understood this in the depths of her heart and mind, but she did not know the details of the process that generated the patterns she observed. Her work identified a new phenomenon of genetic inheritance and provided a scaffold of understanding based on the methods that were available at the time. It was up to those who followed—using more advanced technologies—to fill in the details. But the truth of McClintock's original discovery never wavered, because it was based on solid scientific methods and large amounts of data.

---

[9] B. McClintock, "The Origin and Behavior of Mutable Loci in Maize," *Proceedings of the National Academy of Sciences* 36 (1950): 344–55.

[10] S. B. McGrayne, *Nobel Prize Women in Science: Their Lives, Struggles, and Momentous Discoveries,* 2nd ed. (Carol Publishing, 1998).

[11] R. M. Davison, *American Women Scientists: 23 Inspiring Biographies, 1900–2000* (McFarland, 1999).

Do all ideas proposed to explain natural phenomena eventually become supported by data and accepted as truth? No, not at all. Take, for example, the idea of *flood geology*. A modern "theory" of flood geology was presented in a book published in 1935,[12] and this idea has been reincarnated several times since then.[13] You might think that, just like jumping genes, this was a new idea that could well explain geological features such as the Grand Canyon, so why didn't it catch on? The difference between flood geology and McClintock's jumping genes is twofold. The first is that the proponents of flood geology have to be picky about the evidence they present; they are not adhering to the principles of scientific investigation that require scientists to consider all of the evidence available. The second is that no new corroborative evidence has been uncovered to support the idea of flood geology. There are clear instances where large Pleistocene floods created dramatic geologic and landscape features when ice dams broke (a scenario depicted in the 2006 animated film *Ice Age: The Meltdown*), such as the Columbia River Gorge in Oregon, but such local floods do not support the idea of a worldwide flood. The reason that flood geology is not accepted as science is because there is no evidence to challenge the idea that geological features on our planet are the product of many millions of years of erosion. This idea of *uniformitarianism*[14] was first presented by James Hutton in 1785 and popularized by Charles Lyell in 1830; it has been supported by thousands of scientific investigations and stands as the foundation of modern geological sciences.[15]

The take-home message is that science is not dogmatic; with good data, replication of experiments, and the accumulation of supporting evidence, the scientific community can be convinced that a new model—a new theory—is needed to explain unexpected results. That's exactly what happened with McClintock's jumping genes.

In many ways, Charles Darwin's description and the ultimate widespread acceptance of evolution by natural selection were similar to McClintock's jumping genes. Darwin's case differed because there were many more scholars of the day who seized upon his new theory with great enthusiasm. This was largely because there was growing interest over the previous decades in the idea that species may undergo *transmutation,* a process where individuals of one species change to become a different species. Biological transmutation has its roots in alchemy

---

[12] G. M. Price, *The Modern Flood Theory of Geology* (Fleming H. Revell Company, 1935).

[13] T. Clarey, *Carved in Stone: Geologic Evidence of the Worldwide Flood* (ICR Institute for Creation Research, 2020).

[14] The assumption that the same natural laws and processes we observe here and now have operated everywhere and at all times.

[15] G. H. Scott, "Uniformitarianism, the Uniformity of Nature, and Paleoecology," *New Zealand Journal of Geology and Geophysics* 6 (1963): 510–27.

and the idea that anything can be transmuted from one form to another.[16] It was assumed that all organic and inorganic materials consisted of the same fundamental elements. Alchemy had been practiced for many centuries and was particularly focused on the possibility of transmuting base metals such as lead to noble metals such as gold. The idea was easily extended to the biological world; prior to Darwin, scholars such as Jean-Baptiste Lamarck discussed a similar idea (*transformise*) where characteristics acquired during the life of an individual could be inherited by their offspring. Even Charles Darwin's grandfather, Erasmus Darwin, had discussed the potential for transmutation of one species to another, and it was a popular topic of discussion among natural scholars.

In the years prior to the publication of Darwin's book *On the Origin of Species*, the journalist Robert Chambers anonymously published *Vestiges of the Natural History of Creation*, which pulled together some popular scientific ideas of the day along with transmutation of species. Darwin was not concerned about being scooped by this publication because he recognized that the ideas were poorly supported and the discussion was amateurish. Not long after publication, *Vestiges* was widely criticized by scholars and ultimately discarded. Nevertheless, it was a bestseller and popular in Victorian society. Darwin thought this could pave the way for wide acceptance of his theory of natural selection, which he had been developing for many years and which was finally published fifteen years later in 1859. Darwin's book on evolution proved to be quite popular—the first edition sold out in a single day. While many scholars of the day quickly accepted Darwin's ideas on the origin of species, there were many who favored natural theology over evolution. The dismissal of scientific evidence in favor of religious dogma continues today, with reincarnations of natural theology such as scientific creationism and, more recently, intelligent design.

Darwin's use of the term "evolution" instead of the popular "transmutation" of the day was a well-considered choice. For one, he wanted to establish his theory of evolution by natural selection as a novel idea, not just an extension of previous discussions of transmutation. But more than that, transmutation of species had been viewed as a transitioning from one species to another. Darwin wanted to invoke the notion that evolution was a gradual and ongoing process, that every species has a history and a future, and that evolution is occurring all the time. Darwin's ideas have provided us with a foundation to understand the history of species and the ongoing processes of evolutionary biology that have molded the form, physiology, cellular biology, molecular biology, and genetics of all life on Earth.

---

[16] R. Sharrock, *History of the Propagation and Improvement of Vegetables by the Concurrence of Art and Nature* (1660). Reprinted by Forgotten Books, 2018.

The science of evolutionary biology is about understanding the history of life on Earth and the processes that have generated the amazing diversity of microbes, plants, and animals that we find around us. The tools evolutionary biologists use range from observations of behaviors and measurements of organisms in their natural environments to applications of advanced technologies in cellular and molecular biology and genomics. The data collected is analyzed and interpreted in the context of predictions from a foundation of mathematical theory of evolution that has been continually developed and elaborated since the early twentieth century. Evolutionary biologists have applied advanced mathematical techniques to model the processes of mutation, natural selection, random genetic changes, and gene movement within and among species to generate a robust body of explanatory and predictive theory. The vast amounts of observations, measurements, and genetic data that have been collected have provided tests of hypotheses developed within the framework of evolutionary theory and informed the development and refinement of models that summarize our understanding of evolutionary processes.

Much of our understanding of the patterns and processes of evolution have come from numerous studies of a handful of species. We have focused our efforts on a smaller number of "model" organisms because this has allowed us to develop large amounts of information on each one that can be used to inform the development of more sophisticated experiments and analyses. Early studies were only able to examine aspects of the morphology, anatomy, and physiology of organisms, such as their unique and shared characteristics that are controlled by gene expression and the environment they are exposed to during development. Since the early 1980s, this information has been supplemented by DNA sequencing and information about genomes. Now we know a lot about the biology, genetics, and evolution of humans, mice, fruit flies, the mouse-eared cress, yeast, and more. But advancements in DNA sequencing technologies have greatly reduced costs and improved accessibility, so that now many more species are being thoroughly analyzed. This information is vastly improving our general understanding of the evolutionary patterns and processes that have generated the diversity of life on Earth.

To have a true understanding of the forces that have shaped the evolution of organisms, we need to study them in their natural environments. We know, for example, that the behavior and biology of animals in captivity is different from that of animals in nature. Our best efforts to create natural enclosures in zoos cannot replicate the range of environments and interactions within and among species in their natural habitats. In many ways, humans are no different than animals in zoos; across societies around the world, we have done our best to create artificial environments for ourselves. In industrialized societies, we move from one controlled environment to another as we travel from our homes to our cars to our places of work, education, and play. Those of us who are fortunate enough to enjoy the advantages of modern technologies are constantly buffered from extreme heat or cold. Even as we engage in outdoor activities, we cover ourselves in advanced clothing to keep warm and sunscreen to protect us from the harmful rays of the sun. Many of us have benefited from medical advances that have prolonged our lives and ensured the survival of many individuals who normally would have succumbed to diseases, pathogens, and genetic deficiencies.

Of course, our technological advances have improved our health and prosperity, but the vast infrastructure we surround ourselves with is far removed from the natural environments that shaped our evolution. Many aspects of our biology and behaviors that developed to improve survival in the harsh savannas of eastern Africa have been recast to serve the whims of a self-centered society. We view ourselves as intellectually advanced because we have more capacity for empathy and generosity. In modern society, many of our behaviors appear altruistic, but cooperation and sacrifice for others were favored by selection—otherwise, our ancestors would not have survived. Attempts to understand our evolution from studies of human behavior in our present artificial context will often generate false conclusions because they are not interpreted in the natural habitats where our unique characteristics and behaviors originated.

In this book, we will follow paths supported by evidence and guided by evolutionary principles to develop an understanding of human evolution in the natural habitats of our ancestors. We will examine evidence from human biology and behaviors in the context of the harsh conditions that our ancestors experienced. Our insights into selective forces affecting human morphology and behavior are guided by evolutionary principles and comparison to other species that live under similar conditions as our ancestors. To provide a deeper understanding of what our ancestors endured, we will follow episodes from the life of Launua, a fictional woman who lived some 70,000 years ago. Her experiences provide a more vivid context to improve our understanding of the forces that shaped the forms of our bodies and behaviors. In each chapter we will evaluate evidence and draw conclusions in the context of our ancestors' natural habitat. In the process we will develop an understanding of the cascade of interacting forces that have led to the evolution of our unique appearance and behaviors.

*Summary*—In this chapter we have covered some basic information about how scientists use evidence in the physical world (data, observations) to test hypotheses that were developed within the framework of widely accepted scientific theories. We learned that science is not dogmatic and that with application of the scientific method we can learn new things that change our understanding of how the natural world works. Both Barbara McClintock and Charles Darwin used large numbers of observations to develop new ways of thinking about natural processes, and consequently, they had far-reaching impacts on science and our modern societies. We also began to understand ways of developing conclusions in science based on evidence interpreted using the principles of evolutionary biology. We learned that to understand the selective forces affecting any species, we need to think about how they were influenced by their natural habitats.

# 2

# Life on the savanna

*It was a difficult life for Launua and her clan. The savanna was hot and dusty most of the year and there was little or no rain. They sought out lakes and rivers so they would have a reliable source of water, and they ventured into the savanna to hunt animals and gather roots and grain. The wooded ravine along the shore of the lake was the only environment Launua had known in her short life. She could play in the shallows of the lake or among the trees, but as a member of the clan, she was expected to help out, even as a child. As she worked alongside her mother scraping the bitter-tasting skin from a root, she looked around the camp. There were other children—some were too young to help and toddled around under the watchful eyes of adults or suckled at their mother's breast. Some adults were preparing roots for roasting, just as she was. She watched one of her cousins carrying a digging stick and a basket full of roots as he came down into the ravine from the savanna. It made Launua think about how she wished to run free across the savanna to gather her own roots. But younger*

*Looking Down the Tree*. Mitchell B. Cruzan, Oxford University Press. © Oxford University Press (2025).
DOI: 10.1093/9780197805190.003.0002

*children were not allowed to venture out on their own and could only leave camp with an adult. She enjoyed the excursions when adults led groups of children out to teach them the ways of the land—how to find good roots and grain and how to track animals. She was not paying close attention to her work; her cutting stone slipped and she cut the back of her hand. Before she could cry out, there was a commotion at the other end of camp. Along with many others, she ran to see what was happening. Some clan members were cheering, and many were jumping with happiness to see a group of men and women hunters coming over the lip of the ravine carrying a large animal they had killed. Launua smiled, knowing that now they would have meat to eat for several days. Some clan members ran to help the hunters bring the heavy carcass down the ravine slope. As Launua watched, she was suddenly startled to find an elder standing by her side. Smiling, she handed Launua a digging stick and a basket. Launua looked at the items and then up at the elder with a proud smile. Her wish had come true.*

You can learn a lot about human evolution from books and internet searches. There is plenty of information on the differences between us and our ancestors based on fossils—the typical stooped-over, knuckle-walking ancestor that progresses to an upright and bipedal modern human. The fossil record indicates that our ancestors' skulls became larger to accommodate the increases in the size of their brains and, at the same time, their jaws became smaller as they started cooking their food over fires. We are also learning much about the history of our ancestors and their relatives from analyses of DNA. But what seems to be missing is more information about what drove these changes. The patterns are thoroughly described, but the story feels incomplete.

Evolutionary biologists are generally more interested in the processes of evolution than the patterns; what caused the changes we see in the fossil record? The patterns of change over time tell us a story, but they don't tell us the motivation for the story. In evolutionary biology it is understood that there could be several different causes of genetic changes over generations, and sometimes these changes happen just by chance. But most of the patterns we see in the human fossil record are not random or accidental but directional, so they were most likely driven by natural selection. Natural selection always has a cause, and the source of the cause is often from the environment in which the animals lived. We realize now that to understand the story of human evolution and the forces of selection that led to our unique appearance and behaviors, we need to know more about the ecology of our human ancestors—where and how did they live?

But our direct ancestors were not alone; there were a dozen or more closely related species of humanlike relatives (*hominins*, which include humans, chimpanzees, and bonobos) that lived in Africa during the same time. Science—*evolutionary biology*—tells us that all living things on this planet are related to some degree or another. We often illustrate those relationships with treelike diagrams, where more closely related species are on shorter branches near each other, and distantly related ones are on separate parts of the tree originating from longer branches. The idea that we can group species based on appearance in a hierarchical framework goes back to the 1700s when Carl Linnaeus worked to describe and classify all living things. Most of the major groups that Linnaeus described—including his placement of humans with all the other mammals, and most closely with the chimpanzees and gorillas—have remained the same over the past three centuries. This work continues today, but morphology has been supplemented with vast amounts of information from DNA, and relationships among living species are largely ascertained based on the similarity of *genomes*.[1]

We have DNA evidence for only a couple of our extinct close relatives (archaic hominins including Neanderthals and Denisovans), so our ideas of relatedness among fossil species are based mostly on morphological characteristics; for vertebrates such as fish, reptiles, and mammals including primates, relationships are generally based on evidence from bones and teeth. Using evidence from the fossil record, we have developed diagrams of relationships among our hominin ancestors by comparing the morphological characteristics of their skeletal remains. These diagrams often show connections among species in a branching pattern leading up to modern humans at the top of the tree, or in the case of Figure 2.1, by placing more similar species adjacent to each other. In Figure 2.1, humans are at the top level, and the line extends all the way to the right, indicating that our species is still living today. Other species are shown below humans with horizontal bars that begin when each species first appears in the fossil record and end before they reach the present day, indicating that they went extinct. Looking at the middle of the diagram, we can see that there were at least half a dozen species of our close relatives living around the same time. But now all those "cousin" species are extinct. Think about the dog family (*Canidae*). Domestic dogs, coyotes, and wolves all live together and are so closely related that they can interbreed even though they are different species. As modern humans, we have never known other species that were so similar to us that we might have made friends or even interbred with them. Within the hominins, there were a number of close relatives in the genus *Homo* living around the same time in Africa just a few hundred thousand years ago. Now we are the only one left.

---

[1] The entirety of DNA that makes up the chromosomes in the nucleus of each cell.

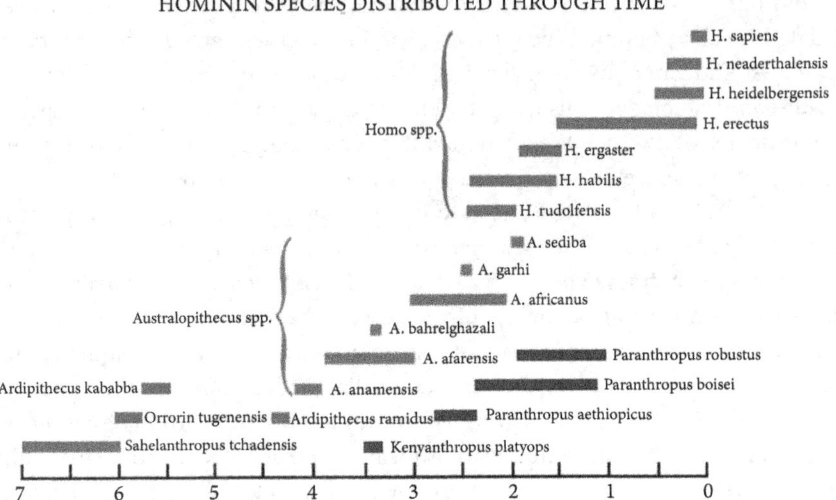


**Figure 2.1** Hominin diversity (excluding chimpanzees and bonobos) over the past seven million years. The horizontal bars indicate the time period that each species was present based on fossil evidence.

Reproduced from Cruithne9 via Wikimedia Commons, Creative Commons Attribution-Share Alike 4.0 International (CC-BY-SA-4.0).

The classic view of human evolution of a knuckle-walking ancestor with a small head and sloping forehead progressing to the upright and bipedal modern human is misleading because it implies that there was a linear progression from apelike ancestors to modern humans. Imagine all of the dozens of hominin fossils that have been found—how do we know which ones are actually our ancestors and which represent side branches that died out? Probably none of the well-known fossils represent our actual ancestors—at least not directly. Of course, they were related to our ancestors, but fossilization is very rare, and finding fossils that are preserved is rarer still. Within the historical lineage of species leading up to humans, there is a single sequence of our ancestors going back through time. Within each ancestral species, of the many thousands of members of that species living at any one time, there is only one sequence of individuals from one generation to the next that we can call our direct ancestors. So the next time you hear claims that a new hominin has been discovered that is our ancestor, you will know better—the newly discovered species is probably not our direct ancestor but more likely another member of the same or a similar species.

You probably have heard of Lucy (a specimen of *Australopithecus afarensis*), the earliest known hominin that was clearly bipedal. When fossils of this species were first found, paleontologists carefully carried them back to their field tent and began piecing them together. As they worked, they played music on an old cassette player. At about the time they were examining the hip bones and beginning to realize their specimen was female, the Beatles song *Lucy in the Sky with Diamonds* was playing, and since then this specimen has been known as Lucy.[2] Even though she is the first known bipedal hominin, Lucy is almost certainly not our ancestor—at least, not our direct ancestor. But she was related to our ancestors; maybe our direct ancestor was someone in her clan, possibly another clan from her same species, or maybe a similar species or *subspecies*.[3] Lucy's fossils, and those of other hominins, tell us about the characteristics that were possible at that time but not necessarily about the appearance of our direct ancestors. Since Lucy was bipedal, it is likely that our ancestors were also bipedal around the same time.

Fossils provide information on when characteristics arose and provide a likely story of the history of relationships among species. But fossils are not enough; to get the full story of our history we need an independent source of information that tells us about relatedness among different species. We need something more than the fossil record showing similar appearances, and that's where DNA comes in.

You probably know about DNA from your biology courses. It's the genetic sequence that makes up chromosomes—the genetic code that makes us who we are and different from each other and other species. You may know about DNA from popular science fiction movies such as *Jurassic Park*, where scientists were able to "reconstruct" extinct dinosaurs based on DNA from blood consumed by fossilized mosquitoes. Knowing the genetic code of some of our recent ancestors has told us much about their appearance, and language and cognitive abilities. But how can we say the information is independent of appearance if genes determine what each species looks like?

You might remember from your biology courses that only a small fraction of genes determines what we look like, and all the genes together represent only a small fraction of all the DNA in our genomes. Most of our genome consists of long stretches of non-functional DNA between genes that is free to accumulate *mutations* (random mistakes in the genetic code), and they happen regularly, providing us with a molecular clock for relatedness—a measure that is independent of the characteristics of each species. Most of the genome does not affect what the individual looks like—it is not affected by natural selection. Over time, mutations gradually accumulate within each species and provide a measure of

---

[2] D. C. Johanson and K. Wong, *Lucy's Legacy: The Quest for Human Origins* (Harmony Books, 2009).

[3] A morphologically distinct group that can interbreed with other subspecies within the same species.

the time since each species split from the others, starting out on its own path of evolution to form a new branch in the tree.

But if mutations are so rare and random, how can we use them as a clock? Let's think about something else that's super rare and random, lightning strikes. The chances of getting a strike in your backyard or your neighborhood are really low. But what if we counted all the lightning strikes in your local region, or the entire state, province, or continent? Don't you think that the number of strikes per year over a large area like that would be pretty constant? Well, it turns out that the average number of lightning strikes per year over large geographic areas is not perfectly consistent, but pretty close. Counting the mutations across the genome is like counting the lightning strikes each year across an entire continent; the number occurring over time is pretty constant. Now we have a tool—a molecular clock—that we can use to estimate relationships among species. Counting up the mutation differences among species over a large proportion of their genomes indicates their relatedness. We can use this information to build a tree—like a family tree for the relationships among species—by noting which species share the most mutations that are not shared by others, we can build a tree, similar to a family tree for the relationships among species—something we call a *molecular phylogeny*.

A molecular phylogeny is much like a tree showing ancestors based on fossils, except that all the branch tips end at the same level across the top (Figure 2.2). Instead of showing only ancestors as we did in the fossil-based tree, the molecular phylogeny shows living species with their names at the ends of the branches. The ancestors of each pair of species are represented as the places where the branches join together (*nodes*) as you go down the tree. In this phylogeny of the great apes, humans are listed as *Homo sapiens*. Next to humans, one branch splits into two branches ending with the names of two species of chimpanzees— the common chimpanzee (*Pan troglodytes*) and the bonobo (*Pan paniscus*). On a longer branch on the other side of humans we find the gorillas (species in the genus *Gorilla*), and then on the longest branch from the base of the tree we see the orangutans (species in the genus *Pongo*). The length of each branch from the node to the tip is based on the genetic distance, which is calculated from mutational differences in DNA representing the molecular clock. The time labels on each node are the estimated ages of common ancestors for the species with branches arising above that point. Since each branching point represents a common ancestor, we can use the age of fossils that correspond to the origins of pairs of branches—ones that have a mix of characteristics from each of the descendent living species—to calibrate the tree (in other words, to translate time based on the molecular clock into years). According to molecular clock differences and the age of known fossils, the split between humans and chimps occurred around

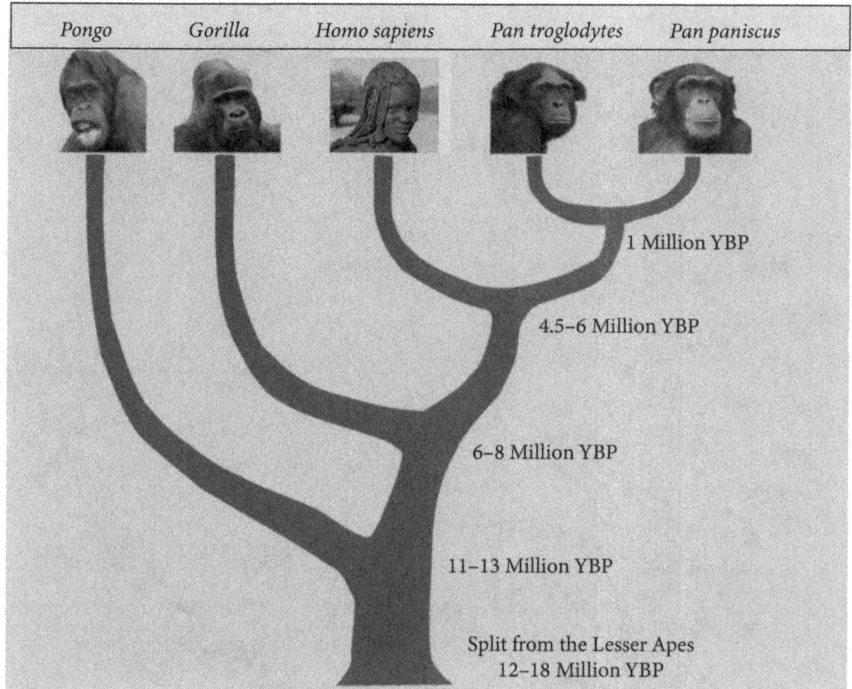


**Figure 2.2**  The phylogeny of great apes. The ages of common ancestors (nodes) are estimated from DNA sequence data and calibrated by fossil evidence.

Adapted from Armin Kübelbeck and HistoriaSalutis (Himba woman), Wikimedia Commons, Creative Commons Attribution-Share Alike 3.0 Unported (CC-BY-SA-3.0).

4.5 to 6 million years ago, and we share a common ancestor with gorillas around 6 to 8 million years ago.

Now, let's zoom out for a moment to see where we are in the wider view of all life on Earth. First, we see a circular phylogeny of all of the mammals, with humans grouped with the other great apes within the primates (Figure 2.3). To the right of the great apes are the Old World monkeys, and to the left are the New World monkeys. Further left are the lemurs and lorises, which are direct descendants of the earliest primates on the longest branches from the base. On the next circular phylogeny, we have zoomed out to see humans (*H. sapiens*) within the group of mammals (under the bracket), which is within the vertebrates, the eukaryotes including all animals and plants, and finally all life on Earth, including the two major groups of single-celled prokaryotic microbes— bacteria and archaea (Figure 2.4). At each step, the branch ending with humans becomes smaller, until it is difficult to distinguish among all the tightly packed

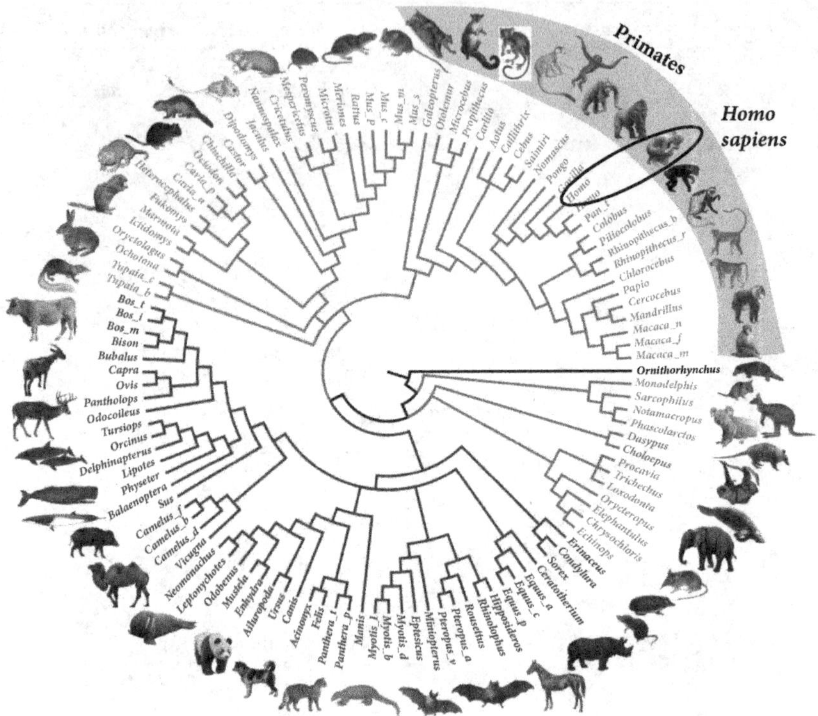


**Figure 2.3** The phylogeny of mammals. The shaded region indicates the primates, with *Homo sapiens* within the oval.

Adapted from OrthoMaM team at the Institute of Evolutionary Sciences of Montpellier (ISEM) and the University of Montpellier (C. Scornavacca et al., *Molecular Biology and Evolution* 36 [2019]: 861–62), Wikimedia Commons, Creative Commons Attribution-Share Alike 4.0 International (CC-BY-SA-4.0).

branches representing myriad microbes and much smaller groups of plants and animals. There are so many branches in the spiral phylogeny for all life on Earth that you have to search hard to even see the primates and the great apes.

Now we can see that all the branches in the tree of life—the molecular phylogeny for all life on Earth—differ in length, from short to very long. Note that there is a continual branching pattern all the way back to the center, to an ancestor we call "LUCA"—the *last universal common ancestor*—which was a single-celled microbe that lived around 3.8 billion years ago. There are no major breaks in this pattern, and the relationships of humans in this phylogeny are similar to any other species. This pattern does not support creation of humans or any other species by a supreme being. It's clear from this molecular phylogeny that all living things on our planet—including humans—are related to each other to some degree or another.

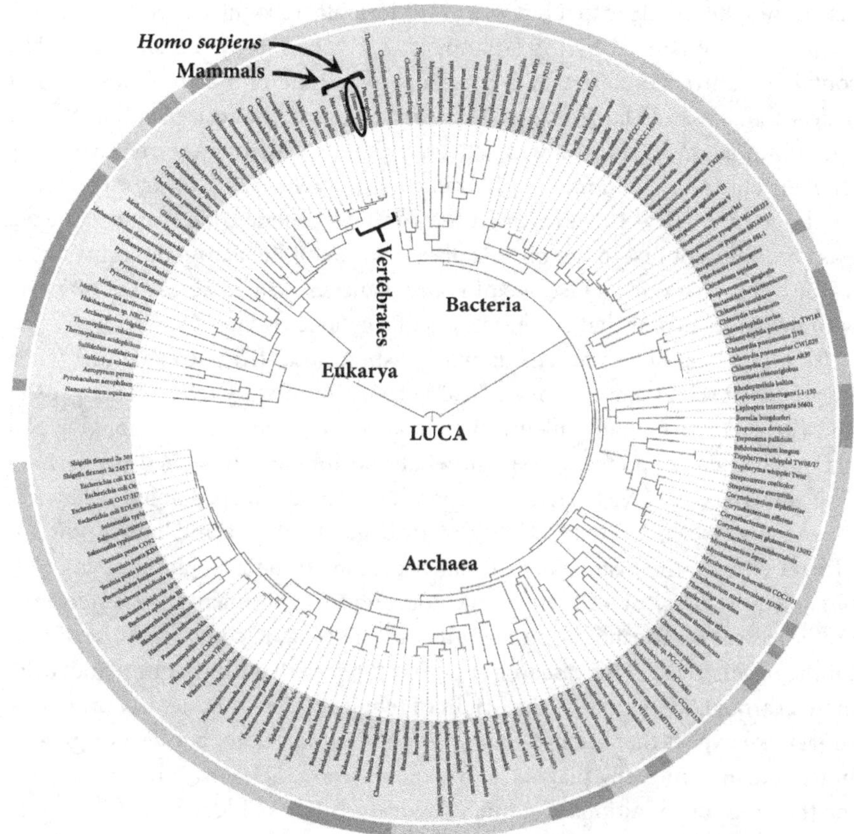


**Figure 2.4**  The phylogeny of all life on Earth. The mammals within the vertebrates are labeled, with *Homo sapiens* within the oval.
Reproduced from Ivica Letunic (2008) via Wikimedia Commons.

But how did it all begin? Life had to get a start on Earth somehow, so where did LUCA come from? One idea, referred to as the *panspermia hypothesis*, is that life exists throughout the universe and is "seeded" as space dust to habitable planets by comets and meteoroids. This idea was popular in the mid-twentieth century as people began to appreciate the complexity of even the simplest living cells and could not imagine how the diverse organic compounds necessary for life could have arisen from inorganic substances. The notion that our species was independently planted here by aliens as well as the origin of fictional humanoids on planets across the universe has been popularized by science fiction movies and series such as *Star Trek* and *Star Wars*. Francis Crick, who was awarded the Nobel Prize for his contributions to our understanding of the structure of DNA along with James Watson (and Rosalind Franklin, who actually discovered the structure first but was never formally recognized for this achievement[4]), was convinced by this argument, and more so by the fact that DNA contains a universal code that is the same for all life on Earth. Crick did not believe that life could have originated or arrived on our planet by random chance but promoted the idea that it was delivered either accidentally or on purpose by an alien entity—a phenomenon he referred to as *directed panspermia.*[5] From Crick's point of view, this explained the complex genetic code that he believed was common not just to life on this planet but to all life across the universe. Is it true, then, that ours is just one of many planets scattered across the universe with living organisms that all share the same genetic code?

It's interesting how many of us will grasp at fantastical explanations when we are faced with seemingly inexplicable facts. But history tells us that gaps in knowledge are eventually filled in by advances in science, and rational explanations gradually emerge to explain what had previously seemed mysterious. This was true of Crick's fallacious proposal. Well before he promoted the idea of directed panspermia, scientists had already begun to understand how conditions present early in the history of our planet promoted the generation of complex organic molecules.[6] Since then, many thousands of experiments have uncovered the processes responsible for generating all the organic compounds necessary for the origin of life. This research supports the idea that life on our planet originated through *abiogenesis*—the origin of living entities from natural processes. In fact, we expect that life not only spawned on this planet but also originated independently on many planets across the universe. Scientists are now exploring the range of conditions that might favor the origin of life in a field of inquiry called *astrobiology.*

But that does not answer the question that puzzled Crick: why does all life on this planet use the same genetic code? It's almost certainly true that life originated many times early in the history of our planet, and at any one time there may have been a diversity of living entities occupying the lakes and oceans. So why don't we see many different types of life today? There are two reasons for this. The first is the consequence of a natural law that was first described by Charles Darwin: any time resources are limited, competition will cull the range of variations among different types of organisms to leave only the strongest. In the case of the earliest living entities, only the ones that were most efficient at obtaining and metabolizing the scarce food available would have persisted. This is the process that Darwin identified as *natural selection,* and if we combine it with a heri-

[4] M. Cobb and N. Comfort, "What Rosalind Franklin Truly Contributed to the Discovery of DNA's Structure," *Nature* 616 (2023): 657–60.

[5] F. H. C. Crick and L. E. Orgel, "Directed Panspermia," *Icarus* 19 (1973): 341–46.

[6] E. T. Parker et al., "Primordial Synthesis of Amines and Amino Acids in a 1958 Miller H2S-Rich Spark Discharge Experiment," *Proceedings of the National Academy of Sciences* 108 (2011): 5526–31.

table genetic code that can transfer improvements from one generation to the next, we have described the process of *evolution by natural selection*. As random mistakes—mutations—occurred in the code, those primitive cells carrying them were eliminated if they made the organism weaker, or they were favored if they improved its ability to obtain food or more efficiently extract energy from food. The organisms possessing these improvements could then grow more quickly and spread to displace the less efficient versions of early microbes.

But early in the history of life on Earth—before LUCA—there was another way organisms could improve their fitness. Once a beneficial mutation appeared in one type of microbe, it could be transferred to other microbes by absorbing DNA that had been released by dying cells. This process of sharing genes among different microbes is called *horizontal transmission* (as opposed to vertical transmission from one generation to the next). Horizontal transmission was important for the evolution of life on Earth leading up to LUCA, and it continues to be important today for the diverse range of microbes inhabiting our planet.[7] But notice that horizontal transmission is an effective mechanism of evolution only if the same genetic code is used by many types of microbes; any that could not accurately translate the information coded by DNA could not benefit from horizontal transmission and would eventually be eliminated as the microbes sharing the same code proliferated. During the time leading up to LUCA, horizontal transmission was prevalent and ultimately responsible for the origin of our microbial ancestors and for the elimination of other life forms that did not share the same genetic code. The emergence of LUCA was a biological singularity, an event that erased all evidence of the diversity of life on Earth that preceded it. There are no descendants of the other diverse types of life that may have occupied early Earth, so we have no way to understand what they may have been like.

While we are not able to appreciate the diversity of life that existed prior to LUCA, we have made great strides to explore the numerous species alive today. Our understanding of the phylogenetic relationships among species is possible because of the vast number of DNA sequencing projects that have been conducted over the last several decades by thousands of research labs at universities and other institutions around the world. This mega-scale effort has resulted in more than one hundred billion nucleotides of sequence data, with hundreds of millions more being added each year. Without really knowing it, we have conducted a grand experiment over the last several decades. Since the late eighteenth century, scholars of the natural world, and later, scientists, have grouped organisms based on their appearance—what traits made them look like the same

---

[7] H. Ochman et al., "Lateral Gene Transfer and the Nature of Bacterial Innovation," *Nature* 405 (2000): 299–304.

species and different from other species. The phylogenies based on DNA did not have to support those early phylogenies, but for the most part they do. The phylogenies we have made based on DNA sequences are nearly the same as those that were made based on physical similarities going all the way back to Linnaeus. I think it is fair to say that the history of life on Earth since LUCA is written in the DNA of all the organisms alive today and has just been waiting there for us to read it.

Now let's focus back on the phylogeny of the great apes. Note that all of the species of great apes share characteristics that make them different from other primates, and they also have characteristics that they inherited from common ancestors, ones that all primates share. Note that all primates have their eyes in the front instead of on the sides of their heads. Most other mammals with eyes in the front for binocular vision—like the cat and dog families—are predators, but primates' diets mostly consist of fruits and vegetation. Binocular vision is important for judging distances and the speed of moving objects, so it makes sense that predators would have both eyes in front. For primates, this characteristic stems from their earliest ancestors that appeared around the end of the Cretaceous before the dinosaurs died out and when the first fruit-bearing flowering plants appeared. In cases like this where almost all members of a group share a characteristic—like fruit eating—the most *parsimonious* (simplest) explanation is that they inherited it from their common ancestor. And in this case of early fruit-eating primates, the molecular phylogeny and the fossil record support this idea. Binocular vision would have been important for these animals because they had to move from branch to branch through trees. Binocular vision would have allowed them to judge distances for leaping from one branch to the next.

Binocular vision and the ability to easily travel through trees in the earliest primates were adaptations for foraging for leaves and fruit. Plants responded to selection pressure from primates and other animals to make immature fruits taste bitter and mature fruits taste sweet and be nutritious.[8] This created an advantage for the plants because their mature seeds could pass through the digestive tracts of animals without being damaged, but developing seeds in unripe fruit would have been destroyed. By eating ripe fruits, animals dispersed seeds to new locations so the plants could spread. At the same time, selection favored individuals of the earliest primates who preferred the sweetest, ripest fruits, because these were the most nutritious. This is a good example of *coevolution* between two groups of species—early primates and plants—that results in benefits to both species: primates have a good source of food and plants have

---

[8] O. Eriksson, "Evolution of Angiosperm Seed Disperser Mutualisms: The Timing of Origins and Their Consequences for Coevolutionary Interactions between Angiosperms and Frugivores," *Biological Reviews* 91 (2016): 168–86.

their seeds dispersed. It also tells us that we inherited our love of sweet fruits and, in general, our "sweet tooth" from our earliest primate ancestors.

Now we can see that some traits are shared among almost all primates, but what characteristics separate the great apes from other primates? If you look at pictures of monkeys, gorillas, chimps, and humans, you'll quickly see that there is one obvious difference; the great apes (and gibbons, which are called lesser apes) don't have tails, but all of the monkeys do. Since this is a trait that all of the members of the group share, we can be pretty sure that the original great ape—the ancestor of all of these species—also did not have a tail. Why would our ancestor lose its tail? With the exception of gibbons, all the other primates have tails, so what is different about the great apes? For one, great apes have much larger bodies than other primates, and they spend much more time on the ground. Natural selection would have favored the loss of a tail as body size increased because a massive muscular tail would have been needed for great apes to be able to hang from branches the way monkeys do. Such a large tail would have been energetically expensive and probably would have interfered with their movement on the ground. Evolution is frugal, not wasteful; any structure that is not useful—that does not increase survival or reproduction—tends to be lost. So that was the fate of the tail of the first great ape, and all descendants of that species—including humans—also lack tails.

You are probably starting to realize that if we think about all the characteristics of any particular species, we can identify some that are inherited from ancestors and others that are unique. Take humans, for example. Most of the great apes spend a good amount of their time foraging in trees, and they use their fists as an extra pair of feet when they are on flat ground. They walk using their feet and the knuckles of their hands because they can't stand upright on their hind legs for very long. But humans are bipedal and are not very good at climbing through trees compared to the other apes. Our ancestors lost those characteristics at some point in the past. As our ancestors became bipedal, their feet became less like hands and more like the feet we have today because selection favored this form for walking and running on flat ground. As we will discuss in the next chapter, bipedalism arose in our ancestors as an adaptation to spending most of their time on the ground instead of climbing through trees. But where did these bipedal ancestors live?

To determine where and how our ancestors lived, we need to focus on the group of our closest relatives. Over the past four to six million years, at least twenty-five species of hominin (excluding chimpanzees) are known from fossil evidence. The older specimens, which lived between four and seven million years ago, include species of *Sahelanthropus, Orrorin,* and *Ardipithecus.* Later species living between about one and four million years ago included *Australopithecus* and

*Paranthropus.* We have fossil evidence of twelve species in the genus *Homo,* the first of which was *Homo habilis,* starting a little more than two million years ago. A number of these species have been identified from fossil evidence only in the last few decades. The relatively high rate of discovery of new species of hominin over the past several decades tells us that we have not yet reached a plateau, so it's likely that there are many other extinct species of hominin that have yet to be discovered and described.

It's often unclear whether a new find represents a new fossil species or is another specimen of a species that has already been described. Note that the way fossil species are defined is based on morphological distinctiveness and requires some degree of guesswork. Consequently, there is not always agreement among paleontologists about which specimens represent a new species or just a variant—a subspecies—within a known species. For example, *Homo erectus* is a highly variable and widespread species, and it's been argued that some of the subspecies actually represent different species. In other cases, descendent species have sometimes been classified as just subspecies of *H. erectus.* Nearly all hominin species and subspecies represent side branches that ended in extinction, and there is only one sequence of ancestors going back through time from modern humans to our shared ancestor with chimpanzees and bonobos. This lineage and sequence of ancestors represents our unique evolutionary history and is the focus of this book.

While the diversity of hominins is intriguing, the majority of species are not our direct ancestors. These species of hominins outside of our direct lineage of ancestors are not of great interest for understanding the evolutionary pressures that led to our unique set of characteristics of how we look and act. So how do we know which of these species are most likely to be our direct ancestors? As we mentioned previously, probably none of them were—at least, not in the sense that you can trace your family tree back in time. But there is a sequence of species known from the fossil record that closely resemble our direct ancestors because they lived in the same habitat and attained similar characteristics around the same time. These close relatives are extremely valuable for us to understand our own evolutionary path and are relevant for us to understand how we attained the unique set of characteristics that define us as humans.

To understand which of the known hominin species are closest to our actual ancestors, we need to consider several lines of evidence, the first of which is the set of characteristics that distinguish each hominin species. We expect that more recent species will be the most humanlike in all their characteristics, and as we go back in time, the specimens will look less like modern humans and more like our closest relatives within the great apes—the chimpanzees and bonobos. One of the primary characteristics that separates us from chimpanzees is the skull, which in chimps is much smaller with a more elongated muzzle and a much lower cranial dome. In addition, their feet, legs, and hips were adapted for arboreal life with four toes that are long and can grasp branches and with the large toe set further back on the foot. Their hands differ from ours with a thumb that is set further back from the fingers and much closer to the wrist, so it is not used for grasping in coordination with the fingers. Our hands have thumbs that are directly opposable to the fingers, so we use both our thumbs and fingers to grasp objects. While the earliest hominin species are much more apelike and were well adapted for an arboreal existence, we are looking for the first species to exhibit more humanlike characteristics and, in particular, the ability to efficiently walk on flat ground—bipedalism. When we scan the available hominin fossils, we find that the first species that was truly bipedal—with feet similar to modern humans—was *A. afarensis* (first known by the specimen called "Lucy"), which lived in eastern Africa in the regions now known as Ethiopia, Kenya, and Tanzania.

The second line of evidence we need to identify the origin of human ancestors is an environment that would have favored bipedal locomotion over the ability to efficiently climb and travel through tree canopies. Recent estimates based on the origins of tree species characteristic of African savannas suggest that this much drier habitat began to displace wet forests in eastern Africa as early as ten to fifteen million years ago,[9] which would have provided ample food resources to be exploited by early hominins. Not surprisingly, the region in eastern Africa where fossils of *A. afarensis* have been found corresponds nicely with the distribution of savanna habitats when they occupied the area around 3.8 million years ago. Another species of *Australopithecus, Australopithecus anamensis*, lived around the same time but occupied wetter and more heavily forested areas in eastern Africa. This species had some characteristics of bipedalism, but wrist and hand anatomy indicate that they were primarily arboreal and engaged in knuckle walking when on the ground.[10] The foot bones of a similar forest-dwelling hominin from around the same time were adapted for grasping branches and not for bipedal locomotion.[11] A somewhat earlier australopithecine species known as *Ardipithecus ramidus* lived in forested regions of eastern Africa around 4.4 million years ago and also had characteristics of bipedalism and arboreality. Although these other species had some characteristics of bipedalism, evidence from the fossil bones available indicate that

[9] T. J. Davies et al., "Savanna Tree Evolutionary Ages Inform the Reconstruction of the Paleoenvironment of Our Hominin Ancestors," *Scientific Reports* 10 (2020): 12430.

[10] B. Richmond, "Evidence That Humans Evolved from a Knuckle-Walking Ancestor," *Nature* 404 (2000): 382–85.

[11] Y. Haile-Selassie et al., "A New Hominin Foot from Ethiopia Shows Multiple Pliocene Bipedal Adaptations," *Nature* 483 (2012): 565–69.

*A. afarensis* (Lucy's species) was the first species that was habitually bipedal, with foot and hand anatomies that were more similar to humans, and probably had diverged from a common ancestor with *A. anamensis* a considerable time earlier than the age of the known fossil specimens. As we will see later, many of the unique characteristics of modern humans have their origins in the hot, dry, and dusty habitats of savannas; consequently, we will focus on *A. afarensis* as the closest model to our first ancestor that was clearly bipedal.

The third line of evidence we need to identify the likely region of the origin of humans comes from analyses of genomic *DNA* sequences. If we take samples of DNA from humans scattered across the globe, we can identify numerous differences that occur at specific locations where one *nucleotide* (i.e., represented by the letters A, C, G, or T) is substituted for another. Each of these differences represents a single mutation that occurred at some point in the past. The newest mutations have not been around very long and so may occur in only a few individuals, and older mutations will be more widespread and may occur in many individuals. We can use these unique and shared differences to produce a genealogy—a family tree—for any arbitrary set of living individuals. Note that this genealogy, known as a *gene genealogy*, is for relationships among individuals within a species or among closely related subspecies, while the molecular phylogeny discussed earlier in this chapter is for relationships among species. If we travel from any living individual up the genealogy, we will see that the individual branches join—they coalesce—two by two into single branches until we reach the top, where we find the single ancestor of all the persons included in the original sample. Note that this common ancestor is based on the sample of people we started with, so the estimate of their *genotype* (the DNA sequence of an individual) will change slightly based on who is included in the sample. We can apply predictions from a discipline known as coalescence theory to our genealogy to tell us two things: how long ago this hypothetical common ancestor lived and where.

Coalescence theory have been applied to trace the ancestry of the human mitochondria genome and the Y chromosome. The first estimates were based on mitochondria, which are derived from an ancient symbiosis between the earliest eukaryotic microbes and bacteria, so mitochondria have their own genomes that are similar to bacteria. Since mitochondria are only inherited through the egg cell and not through sperm, this genealogy represents the ancestry of women going back in time (Figure 2.5). Unfortunately, the scientists who made the first estimate of our hypothetical most recent common ancestors (MRCAs) chose to refer to her as the "Mitochondrial Eve," which gave the impression that this was a real person from our past. In reality, this mitochondrial genotype is hypothetical and represents a large group of ancestors who lived at that time

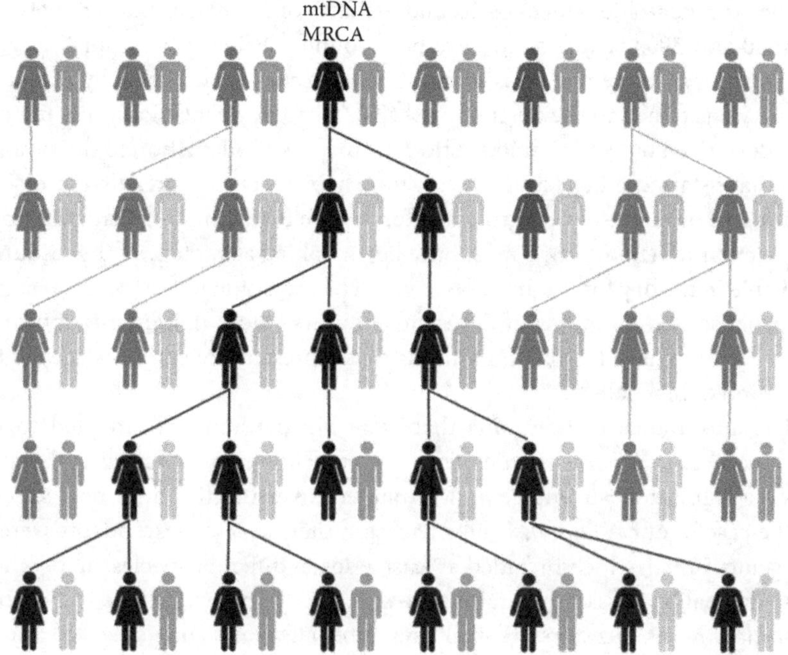


**Figure 2.5** Example of the coalescence estimate of the most recent common ancestor (MRCA) of a group of people alive today based on their mitochondrial genomes.

Reproduced from C. Rottensteiner (2012), Wikimedia Commons, Creative Commons Attribution-Share Alike 3.0 Unported (CC-BY-SA-3.0).

rather than a single individual. Estimates of the age of these common ancestral groups from mitochondria and the Y chromosome date back to 120,000 to 156,000 years ago.[12] More importantly, they identify the same region in eastern Africa that was the home of *A. afarensis* and widespread savanna habitats as the region of origin for all humans.

Now we have three independent lines of evidence that all point to eastern Africa as the cradle of humanity, and we can be confident that our earliest bipedal ancestors lived in dry forest, savanna, and grassland habitats. But note that *A. afarensis* occupied this East African region much earlier than the age estimate of our MRCAs, between 3.9 and 2.9 million years ago. The age estimate for our MRCAs only goes back so far—like a family tree. But we know that our

[12] G. D. Poznik et al., "Sequencing Y Chromosomes Resolves Discrepancy in Time to Common Ancestor of Males Versus Females," *Science* 341 (2013): 562–65.

species is much older since fossils and artifacts of *H. sapiens* that are between 200,000 and 300,000 years old have been found.[13] Prior to that, our lineage is taken up by a different species—probably *H. erectus*—and maybe by *H. habilis* before we get back to the australopithecines like Lucy. Since Lucy already had well-developed bipedal characteristics, the process of adaptation to the savanna must have started much earlier. As savanna habitats began spreading in eastern Africa, one or more splinter groups of forest-dwelling hominins that were probably similar to *A. anamensis* must have begun taking advantage of the resources available in the dry forests and grasslands. The transition to the savanna habitat probably started hundreds of thousands of years before *A. afarensis* appeared, but we do not have fossils of hominins that represent the earliest stages of the evolution of bipedalism.

The appearance of hominins that were bipedal and well adapted to the savanna like *A. afarensis* represented a *key innovation*, a characteristic that provided a significant advantage and promoted diversification into new species. In the case of our *A. afarensis*–like ancestor, their known descendants were in the genus *Homo*, which included at least a dozen different species, all of whom were bipedal but possessed brains that were two to three times the size of their *Australopithecus* ancestors. It's likely that bipedalism and adaptation to savanna habitats along with improved tool-making abilities enabled by the cultural transmission of skills and knowledge (discussed in the final chapter) were responsible for the success of our *Homo* ancestors.

Of particular interest for our human ancestry is a diverse species known as *H. erectus*, which lived at least two million years ago, with the last populations dying out just over 100,000 years ago.[14] At least one subspecies of *H. erectus* occupied eastern Africa and lived in a variety of habitats, including grasslands and savannas. But note that bipedalism was fully developed in our ancestral lineage by this point and apparently allowed our ancestors to be more versatile in their habitat requirements than contemporary species of hominins such as species in the genus *Paranthropus*.[15] *Homo erectus* proliferated well beyond other hominin species and spread across Africa, Europe, and Asia, with the first specimens appearing in China around two million years ago. The origin of *H. erectus* in eastern Africa must have been considerably earlier than indicated by known fossils in eastern Asia because they would have needed time to migrate all the way to China. *Homo erectus* spawned a number of different species of

---

[13] E. M. L. Scerri et al., "Did Our Species Evolve in Subdivided Populations across Africa, and Why Does It Matter?," *Trends in Ecology & Evolution* 33 (2018): 582–94.

[14] Y. Rizal et al., "Last Appearance of Homo erectus at Ngandong, Java, 117,000–108,000 Years Ago," *Nature* 577 (2020): 381–85.

[15] Johanson, D.C., Wong, K. 2009. *Lucy's Legacy: The Quest for Human Origins*. Harmony Books.

*Homo* throughout Eurasia including Neanderthals and Denisovans (possibly via an intermediate species known as *Homo heidelbergensis*)[16] and the diminutive hominin species from the island of Flores (*Homo floresiensis*).[17]

Over the remaining chapters of this book, we will look into our past to understand why we have the particular set of characteristics that we do. We will ask which traits we have inherited from ancestors and which are unique to humans. And most importantly, we will ask how natural selection caused the evolution of these characteristics in our ancestors leading up to modern humans. We will be looking back through our phylogenetic history—we will be looking down the tree.

*Summary*—In this chapter we learned that fossilization is rare, and consequently we have limited representation of hominin diversity over the past 6 million years. It's difficult to use the morphology of fossils to understand which hominin species are closest to our direct ancestors, but we know that the timing of the appearance of different characteristics in known hominins provides information on when our ancestors probably acquired the same traits. We learned how mutations accumulate in divergent lineages, which allows us to estimate relationships among species and to create phylogenies for all life on Earth. We concluded that our species arose from common ancestors within the great apes, and we traced the history of our lineage through the primates, mammals, and vertebrates all the way back to the microbial community known as LUCA that lived some 3.8 billion years ago. From there we looked back in time to the origin of life on Earth and concluded that the occurrence of life on this and other planets is adequately explained via natural processes resulting in abiogenesis.

We discussed how the first phylogenies were based on similarities in the appearances of different species and that the molecular phylogenies developed over the past several decades mostly confirm these original phylogenies. We explored the inheritance of traits and the sharing of similar traits among species. Traits that are shared among most species in the same group are assumed to have first occurred in their common ancestor, and traits that are unique to individual species are more recently derived, most likely through natural selection. To understand how selection favored different traits in our ancestors, we need to know more about how and where they lived.

We developed an understanding that our direct ancestors were associated with dry forest and savanna habitats in eastern Africa. This conclusion was based on

---

[16] M. Dembo et al., "The Evolutionary Relationships and Age of *Homo naledi*: An Assessment Using Dated Bayesian Phylogenetic Methods," *Journal of Human Evolution* 97 (2016): 17–26.

[17] G. D. van den Bergh et al., "*Homo floresiensis*-like Fossils from the Early Middle Pleistocene of Flores," *Nature* 534 (2016): 245–48.

three lines of evidence: the presence of bipedal australopithecines in this region, phylogenetic evidence for the spread of savanna trees several million years prior to the appearance of *A. afarensis*, and gene genealogies placing our *H. sapiens* ancestors in the same region. We learned that species in the genus *Homo* diversified, probably as a result of bipedalism combined with their improved tool-making skills. In particular, *H. erectus* was the first species of humans to spread out of Africa to colonize regions across Eurasia where they diversified into a number of descendent species and subspecies.

# 3
# Walking out of the forest

*Launua's early ancestors lived millions of years before her and in a very different kind of place. They lived in forests that offered a variety of food, and water was easy to find. They spent more of their time climbing through trees searching for sweet fruits than they did on the ground. They looked very different from Launua. Their hands and feet were well suited for grasping branches, and they could easily travel through the canopy, leaping and swinging from one branch to the next. They were less graceful on the ground. They were uncomfortable standing upright, so they rolled their fingers into fists so they could use their hands as an extra pair of feet—walking on their knuckles. Their thumbs were short, lower on their long hands, and away from their fingers, so they would not be able to hold a digging stick as Launua*

*Looking Down the Tree*. Mitchell B. Cruzan, Oxford University Press. © Oxford University Press (2025).
DOI: 10.1093/9780197805190.003.0003

*did, with her thumbs wrapped around one side and her fingers on the other to make a firm grip as she forced the pointed end into the dry soil to unearth roots. They did not have the skills or the finger dexterity to weave reeds to make the basket that Launua used to carry her roots. They would have been uncomfortable and afraid in the open landscape where she worked. With their awkward gait, they would have tired quickly over the long distance that Launua had to run to return to her clan's camp. She kept a sharp eye—peering over the tall grass on the lookout for predators—as she ran lightly over the sunbaked ground with her digging stick in one hand and her other arm tightly wrapped around her basket full of roots.*

Have you ever stopped to think that if our ancestors hadn't began standing on two feet, maybe we would never have invented sports like baseball, tennis, or hockey? And have you ever thought about the fact that if we didn't have such large brains, we wouldn't need noses? It turns out that our appearance, and the appearance of all animals, has been shaped not just by natural selection and random genetic changes but also by limits imposed by *evolutionary constraints*. The appearance of organisms—their morphology, anatomy, and physiology—is not just like soft clay in the hands of natural selection; evolutionary constraints can limit what can be attained and can guide the development of traits over time. There are a couple of different types of evolutionary constraints, and the first one we want to discuss has to do with coordinated changes among different parts of the body.

As we develop from an embryo to an infant and eventually to an adult, a series of developmental programs control the growth of different parts of our body. Sets of genes are turned on in sequence to coordinate the development of different structures from juvenile into adult form. Often, the same program is used during the development of similar parts. The development of a leg and an arm, or a hand and a foot, is similar, so the same sets of programs kick in for similar structures as we grow. These shared developmental programs generate genetic associations (*genetic correlations*) between different structures that cause coordinated responses to natural selection; if there is selection on one trait, then other traits sharing the same programs will respond in similar ways and change over time. This is a good thing because it keeps the body balanced as it grows and as it has changed over evolutionary history. Consider how selection increased the body size of the modern horse from its much smaller ancestors as shown in Figure 3.1. Since all four limbs share the same developmental programs, the constraints imposed by development maintained a balanced animal; as selection increased body size, the lengthening of all four limbs happened in concert.

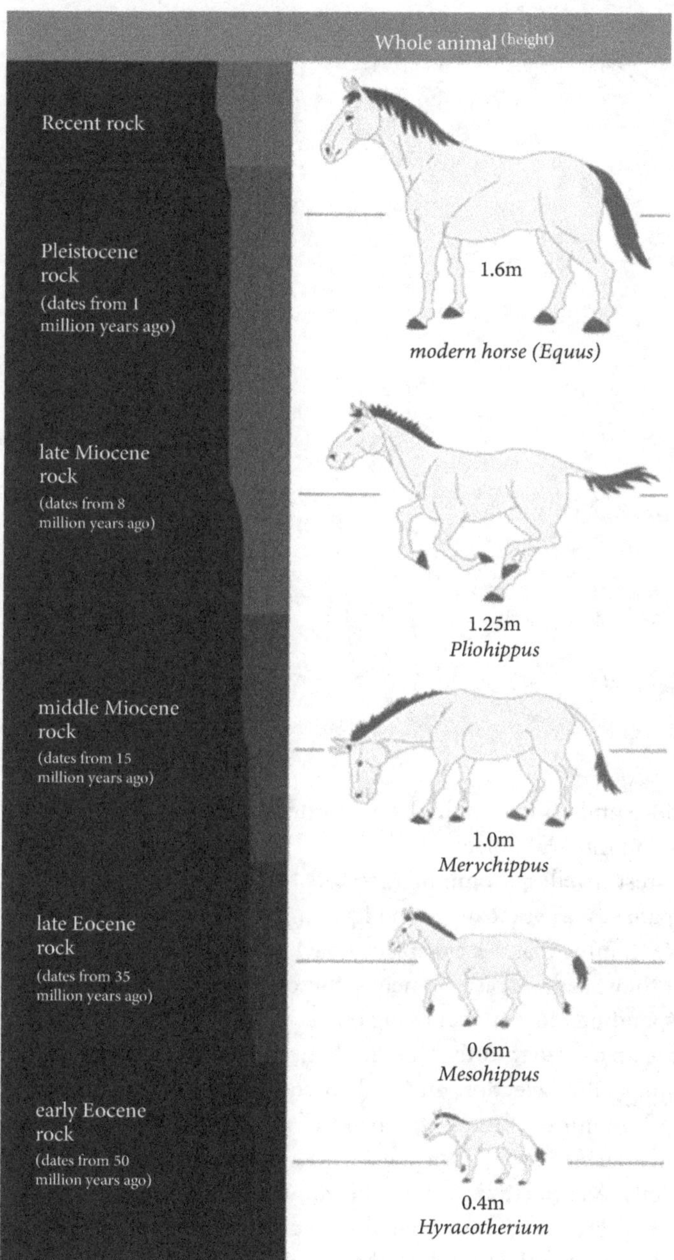


**Figure 3.1** Evolution of the modern horse from an ancestor that lived 50 million years ago.

Adapted from Mcy jerry (2021), Wikimedia Commons, Creative Commons Attribution-Share Alike 3.0 Unported (CC-BY-SA-3.0).

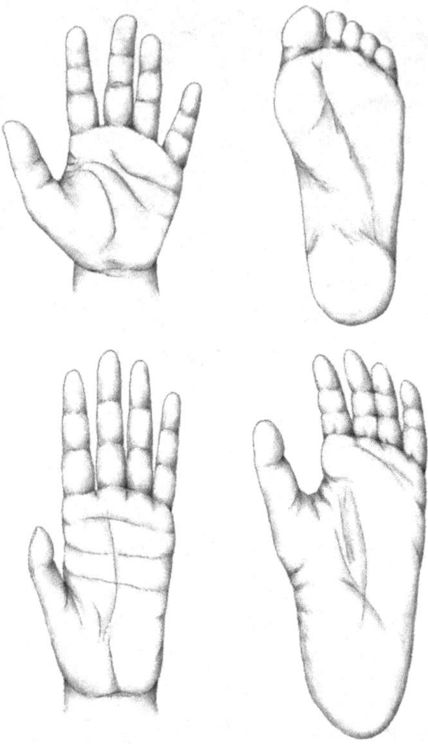

**Figure 3.2** Morphology of human and chimpanzee hands and feet. Katherine Wonder (own work).

But the fore limbs and hind limbs of our hominin ancestors were very different compared to those of horses.

Our forest-dwelling hominin ancestors had feet that were similar to the feet of chimpanzees; as you can see in Figure 3.2 (bottom right), their big toes were more like thumbs that were displaced back toward the heel, and they had long toes that they used to grasp branches. But once our australopithecine ancestors started spending more time moving on flat ground, there was selection to move the big toe up next to the other toes for better balance and more efficient walking and running. But selection on feet had consequences for development of the hands in hominins.[1] Initially, the palm of the hand was much longer and the thumb was farther away from the fingers, as it is in chimpanzees (Figure 3.2, bottom left). As selection favored the elongation of the big toe up toward the other toes, there was a coordinated response in the hands that shortened the palm and elongated the thumb so that it opposed the other fingers.

To understand this, try picking up a drinking glass. Notice how you can firmly grip it by wrapping your fingers around one side with your thumb on the other

[1] C. Rolian et al., "The Coevolution of Human Hands and Feet," *Evolution* 64 (2010): 1558–68.

side. You could do the same with a baseball bat or tennis racket. If you were to hand a bat to a chimp, they could only wrap their fingers around it; they would not be able to get a firm grip using their thumb. That's right—chimps make terrible baseball players! But notice, the same grip needed to hold a bat could be used to hold a cutting tool made from stone, a club, a spear, or a digging stick. Evidence shows that there was strong selection on the bone structure of feet early during the evolution of bipedal australopithecine hominins to elongate the big toe, which simultaneously shortened the palm and elongated the thumb up into a position to oppose the fingers. In other words, the evolution of our ancestors' hands to allow them to tightly grip objects to fashion weapons and tools was a fortuitous accident driven by selection on their feet as they became better adapted for walking on flat ground. Later, selection on hands improved the configuration to optimize them for wielding weapons and tools. But the whole process—the opportunity to refine hands for tool making—was started because of selection on feet for walking across flat ground on two legs (bipedalism).

But there were more evolutionary refinements of hominin hands needed to improve dexterity and the ability to manipulate objects. Try picking up a baseball. Notice how you can hold the ball between your thumb and fingers, and you can exert enormous pressure on the ball's surface. If you are holding a baseball, place your index finger across one seam with your thumb on the opposite seam and your other fingers folded underneath the ball. This is how you throw a curveball. A chimp or a forest-dwelling ancestral hominin could only hold the ball by using their fingers to press it against their palm—they would not be able to throw a curveball, which is another reason chimps would be terrible baseball players.

But our hands are not just adaptations for clubbing and throwing things. We have an amazing ability to manipulate small objects between our fingers and our thumb. Try picking up something very small like a toothpick or sewing needle. Holding and manipulating such small objects require great sensitivity in our finger pads, and we are able to exert a lot of force with our fingertips. Our fingernails are made of the same material—keratin—as the claws of other mammals, but our "claws" are flattened and cover the whole back of the fingertip. Our fingernails provide the support we need for our fingertips to apply force to objects we are manipulating, like snapping our index finger down as we release the baseball to give it the spin it needs to curve. And, as you probably realize, our fingernails are tools in and of themselves, as they are great for prying and peeling.

That makes sense for the evolution of fingernails, but why do we have toenails? There is some speculation that there was selection for toenails to help support the toe tips for bipedal walking and running, or maybe selection favored toenails for protecting our ancestors' toes from damage. Both of these ideas may be valid, but note how similar toenails are to fingernails. It's likely that selection on fingernails resulted in a coordinated response in toenails. Coordinated responses to selection in hands and feet (and other parts of our bodies) have resulted in a lot of similarities between them. There are also smaller differences due to selection that refined different parts of our hands, feet, and limbs to optimize them for different purposes—for example, our toes don't have sensitive pads for feeling and manipulating small objects the way our fingers do.

The seemingly accidental origin of opposable thumbs in our australopithecine ancestors and the refinement by natural selection to produce hands that were adept at tool making are not unique—there are a number of cases in evolutionary biology where novel traits have originated along a circuitous path. An *exaptation* is a trait that originated for one reason but ultimately became an adaptation for a completely different purpose.[2] There are numerous examples of exaptations in evolution, but the most widely cited is the origin of feathers. Feathers first appeared as modified scales in dinosaurs and are thought to have been originally used for thermoregulation and mate attraction.[3] Feathers adorned the fore limbs, heads, and tails of some dinosaurs, but these animals could not fly. Flight developed later as feathered dinosaurs started to glide either by running or by leaping from branches like flying squirrels. But ultimately, evolution refined the early versions of feather-like scales to produce feathers that provide an efficient means for flight in modern birds.

Not all evolutionary constraints are based on developmental programs. Sometimes there are physical constraints that cause evolution to come up with an improvised solution to a problem. If you are a Star Trek fan, you may remember one episode of *Star Trek: The Next Generation* where Captain Picard and several crew members crash-landed a shuttle on a hot, dry desert planet.[4] They realized that the shuttle and food replicators were beyond repair, so they did not have any shelter or water. Their tricorders detected a cave that appeared to have water, but it was a long walk to get there.

"Breathe through your noses," Picard told his crew.

What did Picard mean? To understand why Picard told his crew to breathe through their noses, we need to look at different species of antelopes, including many gazelles and their relatives from Africa (Figure 3.3). One way we can understand how our ancestors adapted to the dry savanna is to examine traits in other species that live in similar habitats. This method, known as

[2] S. J. Gould and E. S. Vrba, "Exaptation—A Missing Term in the Science of Form," *Paleobiology* 8, no. 1 (1982): 4–15.
[3] R. O. Prum and A. H. Brush, "Dinosaur Feathers Came before Birds and Flight," *Scientific American*, May 2014.
[4] *Star Trek: The Next Generation*, episode 83: "Final Mission."

**Figure 3.3**  Saiga antelope. Katherine Wonder (own work).

*comparative biology*, is commonly used to understand the evolutionary origin of different traits. Across species of antelopes, we see a consistent pattern where the species that live in the driest and dustiest places have more nasal surface area. In particular, the saiga pictured in Figure 3.3, which is a type of gazelle that lives in very dry and dusty habitats, has extremely enlarged nasal cavities, so much so that its whole snout looks swollen. The function of this antelope's nose and our ancestors' noses is to filter dust as they inhale and to recover moisture as they exhale—this is an adaptation we see repeatedly in mammals living in dry, dusty places.

But that still doesn't explain why we have noses—why couldn't our ancestors' nasal passages be expanded farther back into their skulls? To answer this question, we need to think about a cross section of the human head. Note that the brain takes up almost all the space in our skulls. The lack of space in the skulls of our early savanna-dwelling ancestors presented a physical constraint that opposed selection to increase the nasal surface area. The only way to accomplish this was to extend the nasal passages outward and downward, from the eyes to the mouth; there was not enough room to expand the nasal passages farther back into the skull because the brain was in the way. The result was a large nose, which we have inherited from our savanna-dwelling ancestors. In contrast, chimpanzees, and almost certainly the forest-dwelling ancestors of bipedal hominins, have very small noses, with the nostrils up close to their eyes—they really don't have noses at all. They do not need a large amount of nasal surface area because they live in wet-forest environments.

Isn't a large nose also consistent with aquatic habitats like those of elephant seals and tapirs, so we can hypothesize that maybe our ancestors spent a lot of time under water?[5] When we examine the evidence for the evolution of different traits, we have to weigh all of the facts that are available. Yes, you can find exceptional cases that seem to point toward different conclusions, but in science, we have to consider all the information; the data leads us on the path to the most logical and simplest explanation. Elephant seals and tapirs have the ability to close their nostrils when they go under water, but of course we don't, and neither did our ancestors. In the case of our noses, there is much more evidence for the association between a large nasal surface area and dry, dusty habitats. The simplest explanation is that during the transition to living in the savanna, our ancestors evolved larger noses. We came to this conclusion using the comparative method, which is a powerful tool for understanding how natural selection has resulted in the evolution of different traits. When we see a consistent association between a particular trait and a type of habitat among species, we can make a strong case for its adaptive value in that environment. In the case of humans, the lack of space in the skulls of our ancestors imposed a physical constraint that caused selection to result in large noses as an adaptation to the dry savanna.

So now you can understand the puzzling statements I made at the beginning of this chapter. We can see the effects of evolutionary constraints on the appearance of modern humans. Genetic correlations due to shared developmental programs cause concerted changes among similar structures; selection on feet for bipedalism caused a correlated response in hands and was responsible for the origin of our opposable thumbs. This change allowed our ancestors to firmly hold objects to use them as tools and gave us the ability to play baseball and other sports. The physical constraint of a large brain was responsible for the origin of our large noses, which effectively filters dust and retains moisture. So, Picard was correct in telling his crew to breathe through their noses so they would be less likely to suffer from dehydration before they could find a source of water.

*Summary*—In this chapter we learned about two different types of evolutionary constraints. The first is based on shared developmental programs that cause

<hr>

[5] E. Morgan, *The Aquatic Ape* (Stein and Day, 1982).

genetic correlations between different structures. We learned that the genetic correlation between foot and hand morphology in hominins is very strong. Consequently, selection on feet as our ancestors became increasingly bipedal caused changes in the morphology of their hands, resulting in the relocation of thumbs that were higher on the wrist and opposite to the fingers. This generated a hand that was adept at grasping using both the fingers and the thumb, whereas our forest-dwelling ancestors were only able to hold onto branches with their fingers. Later, selection refined hand morphology to improve sensitivity of the finger pads, and our ancestors' claws became flattened into fingernails to support the pads. This vastly improved the ability to manipulate small objects.

The second evolutionary constraint we discussed is physical and requires improvisation for evolution to solve a problem. In the case of our earliest savanna-dwelling ancestors, there was strong selection to increase the area of their nasal passages for more effective filtering of dust and retention of moisture. Since there was little available area within the skull, this required extension of the sinuses outward and downward from below the eyes to the mouth, resulting in the nose, which probably first appeared in our australopithecine ancestors who colonized savanna habitats. We noted that chimpanzees, who live in wet forests, do not have sinuses that extend out of their skulls, so they really don't have noses at all.

# 4

# In the heat of the day

*Launua ran at a steady pace across the hot, dry savanna. She had a long run back to camp ahead of her, so she stopped for a few moments in the shade of one of the trees that dotted the landscape to escape the midday sun. She felt hot from the blazing sun, and sweat ran down her forehead, neck, back, and chest. She felt cooler as she ran because the sweat quickly evaporated from her skin, which was covered with thin, short hairs and not the dense fur of her ancestors. When she arrived at her clan's camp, she left the roots she had gathered with some men and women who would prepare them for roasting in the fire. She then headed straight for the shore of the lake to refresh herself in the shallows. Some children were playing in the water, so she joined them to drink her fill and cool off. Launua enjoyed her excursions to harvest roots, but she missed the days when she could play with the other children. It seemed like a long time ago, but she realized she had only been working as a gatherer for a few seasons. In that short time, she had become quite good at digging for roots—she always returned to*

*Looking Down the Tree*. Mitchell B. Cruzan, Oxford University Press. © Oxford University Press (2025).
DOI: 10.1093/9780197805190.003.0004

*camp with a full basket. But she noticed that each day she had to go farther from camp to find good places to dig—her trips had become longer, and she had less time to dig before the heat of the sun became unbearable. Launua's partner and the other men and women who hunted animals had also complained that there was less game, and the clan had not had meat for many days. But the elders were reluctant to leave—this place had been good for the clan for many seasons. They hoped that things would be easier when the rain came.*

The habitats of our human ancestors can tell us a lot about how different characteristics arose and improved their survival. The *savanna hypothesis* is the idea that savanna habitats selected for bipedalism and other characteristics that we see in modern humans. It's clear that the earliest hominins lived in wet forests and spent most of their time in trees; they were knuckle walkers and more similar to chimpanzees than to humans. We know that *Australopithecus afarensis*—Lucy's species—lived in the dry savanna in eastern Africa, and her species must have evolved from earlier savanna-dwelling hominins. It's not clear whether the dry savanna/grassland habitat was constant over a long period as predicted by the savanna hypothesis or alternated with wetter conditions. Either way, we can be confident that there was strong selection for adaptation to hot and dry savanna and grassland habitats because there are a number of characteristics of *Homo sapiens* that are consistent with this scenario, including bipedalism and noses, as discussed in the previous chapter, and the loss of fur, which we will discuss in this chapter.

A sister species to *A. afarensis*, *Australopithecus anamensis* lived primarily in wet forests and had a diet that is similar to modern gorillas.[1] They had some characteristics of bipedalism but probably spent more time in the trees than on the ground,[2] and they were not as well adapted to grassland and savanna habitats as Lucy's species was. There is some discussion about which of these two species is more likely to be closer to our direct ancestor, but as we will see, adaptation to the savanna is more consistent with the characteristics that modern humans have inherited from their ancestors. It is much more likely that our ancestors living at this time were more similar to *A. afarensis* than to *A. anamensis*. The habitats of these two species of *Australopithecus* were very different, so they would have

---

[1] P. S. Ungar et al., "Molar Microwear Textures and the Diets of *Australopithecus anamensis* and *Australopithecus afarensis*," *Philosophical Transactions of the Royal Society B: Biological Sciences* 365 (2010): 3345–54.

[2] G. J. Sawyer, *The Last Human: A Guide to Twenty-Two Species of Extinct Humans* (Yale University Press, 2007).

been unlikely to encounter each other after they had diverged from a common ancestor.

There has been some confusion about the timing of the spread of savannas and dry woodlands in eastern Africa and the appearance of bipedalism in the hominin fossil record. The critical point is that the savanna habitats began spreading ten to fifteen million years ago, and the wet, closed canopy forest habitat of *A. anamensis* was shrinking by at least four million years ago. We know this because we can date the spread of dry forest in eastern Africa using molecular phylogenies for the tree species that characterize this habitat today.[3] The change was very slow and unnoticeable during the lifetime of an individual. As the wet forest shrank, one or more clans of the common ancestor between *A. anamensis* and *A. afarensis* began taking advantage of the food that was available in the savanna, and Lucy's species arose from the colonists of the dry forests and grasslands. The evolution of the more advanced characteristics of bipedalism appears to correspond well with the transition to dryer forest and savanna habitats, but how?

There are several different ideas about why early humans became bipedal, and not all of them are well supported by facts or evolutionary logic. One idea is called the *arboreal fruit-gathering hypothesis*, which suggests that selection favored bipedalism in the savanna for foraging for fruit from trees while standing on the ground or on lower branches to reach up into the canopy. But this really makes no sense, because the savanna trees in Africa do not produce edible fruits. It's possible that foraging for fruit in wet forests was the selective force that favored some characteristics of bipedalism in *A. anamensis*, or perhaps the common ancestor between *A. anamensis* and *A. afarensis*. If this were the case, then the first colonists of the savanna would have had some advantage of preadaptation for bipedalism in their new habitat. Bipedalism in Lucy's species and modern humans is consistent with selection to move efficiently over long distances on flat ground, so even if the earliest savanna inhabitants had a bit of a head start, the characteristics of bipedalism were substantially refined by the time *A. afarensis* appeared.

Some ideas about the evolution of bipedalism sound pretty good at first, but not when you think about them more carefully—for example, the hypothesis that early humans "learned" how to be bipedal while wading in the water, perhaps to catch fish. Chimpanzees are deathly afraid of water and cannot swim, but humans love swimming and all sorts of water sports. We know that our ancestors lived near lakes and rivers, so they probably spent a lot of time in the water. But is wading in the water helpful to achieve the balance necessary to be bipedal?

---

[3] T. J. Davies et al., "Savanna Tree Evolutionary Ages Inform the Reconstruction of the Paleoenvironment of Our Hominin Ancestors," *Scientific Reports* 10 (2020): 12430.

Not really. Chimps can stand upright and run for short distances on two legs, so it's easy to imagine how selection could slowly improve bipedalism starting with our forest-dwelling ancestor. In the previous chapter we discussed how the forces of selection that shaped our feet must have come about from walking and running on flat ground, not wading in the water. While we cannot conduct experiments to test some of the ideas about human evolution, we can make conclusions based on the information we have and our knowledge of evolutionary processes. As mentioned above, the first savanna inhabitants may have had some bipedal ability, which would have made the transition to savanna habitats a bit easier.

The logic of evolution tells us that selection will favor any characteristic that improves survival and reproduction. For our ancestors on the savanna, that meant being able to travel long distances without getting too hot while carrying the things they needed. The original proponents of the savanna hypothesis have also suggested that bipedalism would have been advantageous in watching for predators in grasslands—like meerkats or prairie dogs standing on their hind legs. This seems reasonable, but it was probably a secondary advantage of bipedalism; selection for efficient locomotion across flat ground was most likely the primary force favoring bipedalism.

The savanna was vast and food was available, but it was spread out over large areas, so survival required traveling over long distances. Forest-dwelling australopithecines were probably much like chimpanzees; they could move very quickly over short distances on flat ground but quickly tired over longer distances. Our savanna-dwelling australopithecine ancestors, similar to Lucy, were bipedal, but their feet were adapted more for walking than running.[4] It is not until later in the genus *Homo* that we see adaptations for running, including an upright torso and neck, shorter forearms, a wider spine to absorb impact, longer legs, shorter toes, and tendons in the ankles and feet to absorb impact and provide a stiff arch. The observation that savanna-dwelling australopithecine were better adapted for walking than running is consistent with the conclusions from nitrogen isotope analyses indicating that they were primarily vegetarian[5]. The animals that the savanna-dwelling hominins hunted would have been difficult to catch in a sprint because they generally could run much faster over short distances. The evolution of running ability in later *Homo* species probably came about as hunting and meat consumption became more common. We know that modern Indigenous peoples in Africa and Australia run long distances to chase

---

[4] D. M. Bramble and D. E. Lieberman, "Endurance Running and the Evolution of *Homo*," *Nature* 432 (2004): 345–52.

[5] Lüdecke, T., et al., "*Australopithecus* at Sterkfontein did not consume substantial mammalian meat," *Science* 387 (2025): 309.

an animal until it becomes exhausted and can be easily killed.[6] It is reasonable to assume that our bipedal ancestors did the same thing, an idea referred to as the *endurance running hypothesis*. The ability to run efficiently over long distances would have been critical for survival on the savanna; however, some have disputed this hypothesis and pointed out that improved running ability may have been a side effect of selection for more efficient walking.[7] Nevertheless, it's probably safe to say that becoming bipedal was not just about avoiding predators but also about being one.

Other hypotheses for the origin of bipedalism and its improvement also make sense—for example, the idea that bipedal walking over long distances across flat ground was more efficient than knuckle walking. This seems reasonable because human ancestors were thought to be mostly nomadic—moving around as local food sources became depleted. Selection would have favored the ability to travel easily to find good sources of food. It also would have freed up their hands so they could carry tools and weapons, which were valuable considering the time and effort it would have taken to make them. The ability to carry things while traveling is an advantage of bipedalism in later *Homo* species, but it is unlikely to be one of the primary drivers for the transition to bipedal locomotion because australopithecines probably did not have much if any tool-making skills. The forest-dwelling australopithecine ancestors of *A. afarensis* did not have the adaptations in their hands needed to fashion complex tools. Even though Lucy's grasping hands would have been more adept at manipulating objects, it's unlikely that her species had the cognitive skills for tool making much beyond what modern chimpanzees can do. Their hands had an intermediate morphology between chimpanzees and modern humans.[8] No modified stones or bones associated with this species have been found, but they may have used naturally formed objects as tools.[9] We also know that modern chimpanzees and primates have the capacity to fashion simple tools,[10] so it's not unreasonable to assume that our australopithecine ancestors did as well. So, it appears that the ability to carry things was a secondary advantage that probably came about later with the appearance of species of *Homo*. Available evidence indicates that *Homo erectus* had hands and feet that were similar to modern humans, so it is likely that earlier

[6] E. Morin and B. Winterhalder, "Ethnography and Ethnohistory Support the Efficiency of Hunting through Endurance Running in Humans," *Nature Human Behaviour* 8 (2024): 1065–75.

[7] K. P. Deckers, "These Bones Were Made for Jogging: An Analysis of the Lower Limb Skeletal Evidence for the Endurance Running Hypothesis," *Inter-Section* 3 (2017): 7–13.

[8] C. Rolian et al., "The Coevolution of Human Hands and Feet," *Evolution* 64 (2010): 1558–68.

[9] S. P. McPherron et al., "Evidence for Stone-Tool-Assisted Consumption of Animal Tissues before 3.39 Million Years Ago at Dikika, Ethiopia," *Nature* 466 (2010): 857.

[10] C. Boesch and H. Boesch-Achermann, *The Chimpanzees of the Taï Forest: Behavioural Ecology and Evolution* (Oxford University Press, 2000).

species of *Homo* leading up to *H. erectus* became increasingly skilled at making things with their hands.

Another major advantage for the evolution of bipedalism in our ancestors is the idea that walking upright would mean staying cooler on the hot savanna. For one, bipedal savanna inhabitants would have less skin exposed to the direct sun than knuckle walkers, who walked stooped over. I think we all know how hot pavement can get in the summer—the sunbaked soil of the savanna would have been very much the same. The air temperature is always cooler above the sunbaked ground, so standing upright and getting exposure to cooling breezes would have helped keep them from getting too hot while traveling on the savanna. This *thermoregulation hypothesis* for the evolution of bipedalism assumes that selection would have favored individuals who could conserve water by staying cooler and who would therefore have been less subject to heat exhaustion. While selection for efficient locomotion across flat ground was the primary driver for the evolution of bipedalism, it's probably safe to say that improved thermoregulation also favored an upright posture.

The hot climate of the savanna prioritized many characteristics that would have helped the earliest australopithecine inhabitants to remain cool. We know that our ancestors lost their thick fur and developed thin, short hairs that made them appear hairless at some point in our history, so we have two questions: why and when? Did it happen before or after our ancestors became adapted to the savanna? And what were the selection pressures that caused the loss of hair?

The most widely accepted idea for the loss of fur in our ancestors is that the combination of bare skin and sweat glands allowed them to cool themselves more efficiently. Humans have about the same number of hair follicles as chimpanzees, but humans have very fine, short hairs, so their bare skin is mostly exposed. We have more sweat glands than any other mammal; in fact, most mammals don't sweat at all. Mammals with lots of fur cool themselves by panting or using small blood vessels in their extremities for heat exchange. For example, rabbits and their close relatives use their ears to radiate heat; jackrabbits live in hot and dry desert habitats and have long ears, while their close relative, the pika, lives in cold climates at high elevations and has very short ears to reduce the loss of body heat. What about other mammals that don't have much hair? Elephants, pigs, and hippos don't have sweat glands, but they have other ways to keep cool. Elephants cover themselves in mud, hippos wade in the water, and pigs pant. That's right, pigs don't sweat—at least not very much. So when you say, "I'm sweating like a pig," it really means you're not sweating much at all.

What about the *aquatic ape hypothesis* to explain the loss of hair? This is the idea that at some point our ancestors spent a lot of time in the water, so perhaps

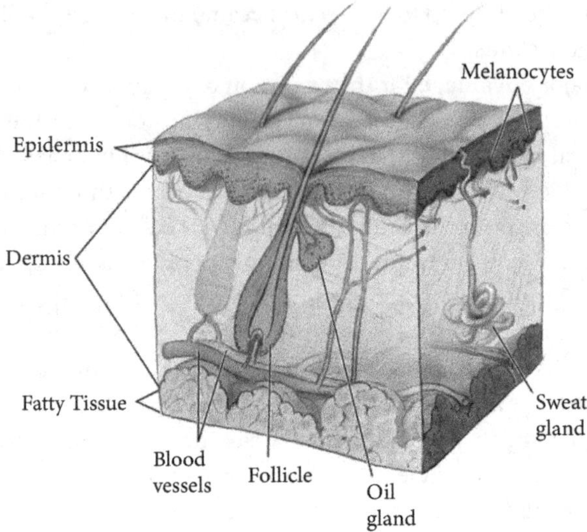


**Figure 4.1** Human epidermis cross section.

Reproduced from Don Bliss (2010), National Institutes of Health
via Wikimedia Commons.

they would have lost their fur, like marine mammals such as whales and dolphins
and freshwater mammals such as hippopotamuses did.[11] Proponents of this idea
suggest that our ancestors became less hairy as they adapted to spending a lot of
time in an aquatic environment. They point out how humans seem to have an
affinity for water and swimming, and even our infants can swim at a very young
age. As we mentioned before, chimpanzees never go in water—not even to wade
in the shallows.

It seems like a pretty good idea, right? But is all the information we have
consistent with the aquatic ape hypothesis? Not really. In fact, the aquatic ape
hypothesis as an explanation for hairlessness in humans has been widely rejected
by scientists, and we can see why by examining the evidence. In the great apes,
there is a correspondence between the absence of fur and the presence of sweat
glands—chimpanzees only have sweat glands on their palms, which is one the
only places where they don't have hair. But the skin of humans is unique because
it is much thinner than other hairless mammals (Figure 4.1). In fact, we have the
thinnest skin of any mammal, and there is a high density of blood vessels just
beneath our skin; thin skin covered in sweat and lots of blood vessels leads to
efficient heat loss.

[11] P. Rhys-Evans, *The Waterside Ape: An Alternative Account of Human Evolution* (CRC Press,
2019).

In the aquatic ape hypothesis, the explanation for thin skin is that it allowed our ancestors to cool off quickly when they were in the water, so they could stay cool just like hippos. But let's look at other differences between hippos and humans. Hippos and other aquatic mammals have large amounts of fat that insulate them from the temperature of the water. But humans don't have that layer of fat, and we have difficulty thermoregulating when we're in water that is too cold or too hot; we can't stay in cold water for long periods of time without a wetsuit. Also, not all aquatic mammals are hairless. Sea otters and river otters, for example, spend most of their time in the water and have a thick coat of fur.

Having sweat glands distributed over most of the body and thin, hairless skin underlain by lots of blood vessels would have been an advantage for early hominins living on the hot savanna. When we sweat, we cool ourselves via evaporative cooling. You might remember from your chemistry class that water has the highest specific heat of any common liquid, which means that evaporation of water is an effective way to cool off. You may know that we sometimes use evaporation to cool our homes and buildings. Swamp coolers work by passing air over panels of dripping water and blowing the cool air into our homes to keep them comfortable in hot weather. But the process of evaporation actually makes the water much colder than the air. Cooling towers take advantage of this by circulating the cold water for air conditioning systems. In the same way, we can efficiently lose heat from our bodies by evaporating sweat so that our skin becomes much colder than the surrounding air. Think about the last time you got out of the water on a hot day—how cool it felt when your skin was wet. If we were furry, our sweat would only get the fur wet; evaporation would only cool our hair and would not be effective in cooling our skin. The fur would also shield our skin from the drying air, so the rate of evaporation, and consequently its cooling effectiveness, would be much lower. When your head hair gets wet from sweat, it's just annoying, and not helpful in keeping you cool. It is the combination of hairlessness over most of the body, sweat glands, thin skin, and a high density of blood vessels just beneath the skin that enables effective cooling and that greatly improved the survival of our ancestors on the hot and dry savanna. But that does not answer the question of *when* our ancestors became hairless.

There is a way we can figure out when our ancestors lost their hair—we can ask the lice![12] Looking at Figure 4.2, which shows a molecular phylogenetic tree for lice that parasitize humans and gorillas, we see that there are four species of lice across the top; three are human lice and one is a louse that is found only on gorillas. The first thing we notice is that the split between the head louse and the body louse in humans occurred about 100,000 to 110,000 years ago. Lice use their claws to grip hair or, in the case of the body louse, threads of clothing. This

[12] R. A. Weiss, "Apes, Lice and Prehistory," *Journal of Biology* 8 (2009): 20.

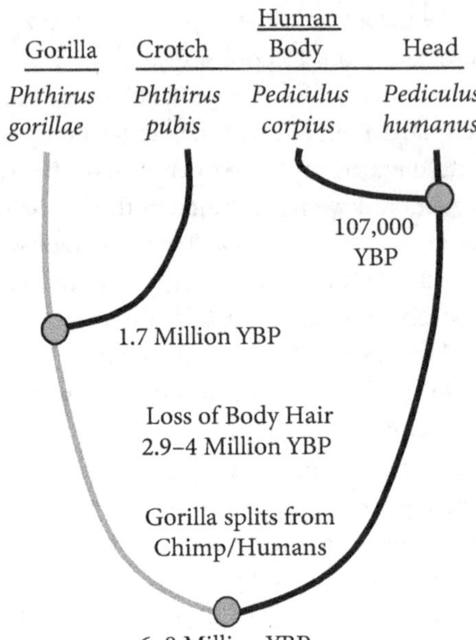


**Figure 4.2** The phylogeny of human and gorilla lice. M.B. Cruzan (own work).

date helps us understand when our ancestors started wearing clothing; the claws on body lice are smaller to match the smaller diameter of the threads in woven cloth. Selection favored smaller claws as our ancestors switched from wearing clothing made of skins to clothing made of woven fabric. Now if we turn our attention to the other two species of lice—the human pubic louse and the gorilla louse—we can see that they are more closely related to each other than they are to the other two lice that parasitize humans. The phylogeny shows the split between the head louse and the gorilla louse at about the same time humans and gorillas shared a common ancestor 6 to 8 million years ago. But the phylogeny also shows a shorter branch—the human pubic louse—off the gorilla louse branch around 1.7 million years ago, which is when our ancestors became infested with gorilla lice in their pubic areas. These lice have very large claws; it turns out that gorilla hair and human pubic hair are about the same thickness, so this was an easy transition for those lice to make.

Yes, that's right; we got "crabs" from gorillas. But note the timing—it means that our ancestors were hairless over most of their bodies by the time they acquired pubic lice. In fact, it makes sense that our hominin ancestors lost hair and increased the number of sweat glands as they transitioned to the savanna habitat, and at about the same time they became bipedal, which must have been earlier than 4 million years ago. Lucy's species lived between 3.9 and

2.9 million years ago, so we can make a strong argument that Lucy and other savanna-dwelling hominins were nearly or completely hairless.

You might be wondering right now why we have pubic hair and head hair—what's up with that? The other great apes do not have curly pubic hair like humans do. In fact, they actually have less hair in their pubic regions. There has been some speculation that pubic hair appeared after our ancestors became mostly hairless and served as a visual signal of sexual maturity.[13] But this idea does not make much sense when you realize that our hominin ancestors in eastern Africa lived close to the equator and consequently had black skin; dark hair on dark skin does not serve as much of a visual signal.

A more likely explanation is that pubic hair serves as a mechanism of chemical signaling. If you look at a diagram of a cross section of skin with pubic hair, you will see the epidermis, dermis, and fat layers along with hair follicles and apocrine glands (Figure 4.1). These glands are only associated with hair follicles, particularly with armpit and pubic hair, and are sometimes called *apocrine sweat glands*. But the apocrine glands in our pubic and armpit regions produce an oily secretion instead of watery sweat; they contract and excrete liquid in response to emotional stress and release volatile compounds—pheromones—that are thought to be important for the subconscious communication of fear or sexual arousal.

Why do you think our hair in these regions is stiff and curly? We don't know for sure, but it's likely that this type of hair is more effective at volatilizing apocrine gland secretions because the curls hold droplets out away from the skin. In other words, pubic hair is probably an adaptation for chemical communication by releasing pheromones, which have subconscious effects.[14] These ideas are somewhat speculative, but it makes sense that these characteristics are an adaptation for the release of pheromones given the strong association between specialized glands and pubic hair. So maybe think about that when you're deciding how much and where to shave or have hair removed.

We know much more about the adaptive value of head hair. Head hair for men and women is effectively the same, but there are cultural reasons that men and women tend to wear their hair differently. You probably know that the best way to keep warm in cold weather is to wear a wool hat because we lose the most heat from our heads and that we can stay cooler in the hot sun by shading our heads with broad-brimmed hats. Selection retained head hair in our ancestors because it was an adaptation for thermoregulation. Our brains are sensitive to excessive

---

[13] P. B. Gray and J. R. Garcia, *Evolution and Human Sexual Behavior* (Harvard University Press, 2013).

[14] Verhaeghe J, et al., "Pheromones and their effect on women's mood and sexuality," *Facts Views Vis Obgyn*, (2013) 5:189–95.

temperatures—hot or cold—and head hair helps keep us cool in the hot sun and warm in cold weather.

Now we can understand that bipedalism, loss of fur, thin skin, and sweat glands originally arose as adaptations to life on the hot, dry savanna and occurred very early in the first savanna-dwelling australopithecine hominins that pre-dated *A. afarensis.* By the time Lucy's species appeared, hominins in our lineage were most likely bipedal for more efficient travel over flat ground. This mode of moving had the benefit of allowing them to look out for predators and to carry their possessions in their hands. When they became bipedal, or soon thereafter, our hominin ancestors became mostly hairless. Their thin skin and sweat glands over most of their bodies allowed them to maintain cooler body temperatures in the hot and dry forests and savannas of eastern Africa.

*Summary*—We established the idea that our australopithecine ancestors had been subjected to strong selection for adaptation to life on the savanna. Bipedalism most likely originated as an adaptation for efficient locomotion over long distances across flat ground and had the added benefits of improved thermoregulation, predator avoidance in grasslands, and the ability to carry things while traveling. We discussed the loss of fur as an adaptation for cooling efficiency; the combination of loss of fur, the spread of sweat glands over most of the body, and a high density of blood vessels just beneath very thin skin provided effective cooling by evaporation of water from the skin's surface. The phylogeny of lice associated with humans and gorillas is consistent with the loss of fur during the time when our ancestors first became adapted to savanna environments. We noted that head hair was important for buffering the increasingly large brain from extreme temperatures, and the thick, curly nature of pubic and armpit hair and its association with apocrine glands are consistent with the hypothesis that these are adaptations for the volatilization of pheromones.

# 5

# A new life

*The decision was finally made to leave the only place that Launua had known all her young life. Some of the hunters who had ventured farther said that there was a good place at the far end of the lake where the water flowed into a river. They all worked together to gather food and their most valuable belongings—the tools they had fashioned from wood, bone, and stone. The skins and poles they used for shelters became sleds that they dragged to carry supplies and some of the youngest children. It would take them many days to reach the new place the hunters had found. During the trip there was much to care for, including infants and baskets full of roots and grain—everyone carried something in their hands or arms as they traveled across the savanna. The clan traveled for several days. Their progress was slow and the path was difficult. They had to travel overland, away from the lake, to avoid the steep cliffs along the shore. They carried water with them in bags made from animal skins, but it was not enough. On the third day the wind blew dust, so they tried to cover their faces with*

*Looking Down the Tree.* Mitchell B. Cruzan, Oxford University Press. © Oxford University Press (2025).
DOI: 10.1093/9780197805190.003.0005

*their arms as they walked, and everyone was thirsty. One infant and an adult who had been ill died along the way. They had to leave the bodies for scavengers—the survival of the clan depended on reaching the next camp.*

*As Launua walked, she thought of her partner. They had been sharing the same sleeping place for many days now, and she knew that women who shared beds with men often had babies. She did not think she would have a baby because her breasts were not full of milk— all of the women with babies had milk. But she noticed that her belly seemed to be larger than before, and something else; her bleeding during the time of the full moon had stopped. She was worried because she had seen other women die during childbirth—sometimes the baby survived and was cared for by others, and sometimes not. She longed to be with her partner—to be comforted and reassured that everything would be fine—but she could not because he was walking ahead of the mothers, caretakers, and children with other men and women to keep watch for predators. When they finally reached the place for a new camp, they dropped their things and went to the river that flowed from the lake. As Launua waded into the water, she felt renewed. She had seen several good places to dig for roots near their new camp, so she was encouraged that food would be easier to find here. Her partner joined her and they started a splashing game. She was happy to be with him and she looked forward to their new life in this wonderful place.*

*Launua and her clan thrived in their new home. On their first hunt, the men and women in the hunting party brought in a large animal to skin and cook over their fires. Launua and the other gatherers also had good luck digging for roots. They tasted the roots of a plant they had not seen before and found that it was good, and it seemed to be common in the local area. But Launua was worried. Her belly had continued to grow and it was getting difficult for her to run across the grassland as easily as she used to. The older clan members had noticed and started to comfort and encourage her. She was grateful that her partner had continued to spend time with her—their intimate moments late at night when the camp was quiet were very comforting for her. Not so many days later, Launua felt a sharp pain while she was digging with the others. She cried out and curled into a ball. One of the older men helped her back to camp; they stopped along the way each time the pain came and then continued as quickly as they could. At one point, water gushed from her, and her companion looked around nervously to make sure there were no predators nearby. They finally made it to camp, and the rest of the day was just a blur for Launua—after so much pain and effort, she was exhausted. Finally, the baby came, as well as the afterbirth. Some elder women and men were there to help and seemed to know what to do—they cut the cord and used coals from the fire to sear the end. They did the same to the tear that was bleeding where the baby had come out of Launua's body—birthing the baby's head had been the most difficult and painful part. But now she lay calmly with the baby on her chest, and her milk was beginning to flow. The baby feeding at her breast gave her pleasure and she felt relaxed—she was full of hope for her new family's future.*

We know that humans have large heads because we have big brains, but has it ever occurred to you that our ancestors with their large brains might not have survived if not for the fact that they were very inbred and lived in clans? If you look at Figure 5.1, which shows the historical sequence of the skulls of our hominin ancestors, the first thing you might notice is that brain volume generally increased from the earliest to the most recent. Lucy's (*Australopithecus*) skull volume was less than a third of the size of modern humans and only slightly larger than a chimpanzee's skull. Recent evidence suggests that *Australopithecus afarensis* used the sharp edges of stones to butcher animals,[1] but there is no evidence that they fashioned tools, and no stones or bones that show evidence of modification have been found associated with this species. In a sense, Lucy's species was not much more than a bipedal chimpanzee. A more recent species, *Homo erectus*, had a brain volume that was twice the size of Lucy's and about two-thirds the size of modern humans. Neanderthals actually had slightly larger brains than humans, but that does not mean that they were smarter than us. Rather, it may have been an adaptation to living in a cold climate because a larger head would have less surface area to total volume, so it would lose less heat than a smaller one. The large differences in brain size we see among hominins indicate general cognitive ability, but smaller differences within species do not. We know, for example, that brain size in modern humans varies quite a bit and is only a weak predictor of intelligence[2]—you might be surprised to learn that Albert Einstein had a very small brain! The main

---

[1] S. P. McPherron et al., "Evidence for Stone-Tool-Assisted Consumption of Animal Tissues before 3.39 Million Years Ago at Dikika, Ethiopia," *Nature* 466 (2010): 857.
[2] C. Koch, "Does Brain Size Matter?," *Scientific American*, 2016.

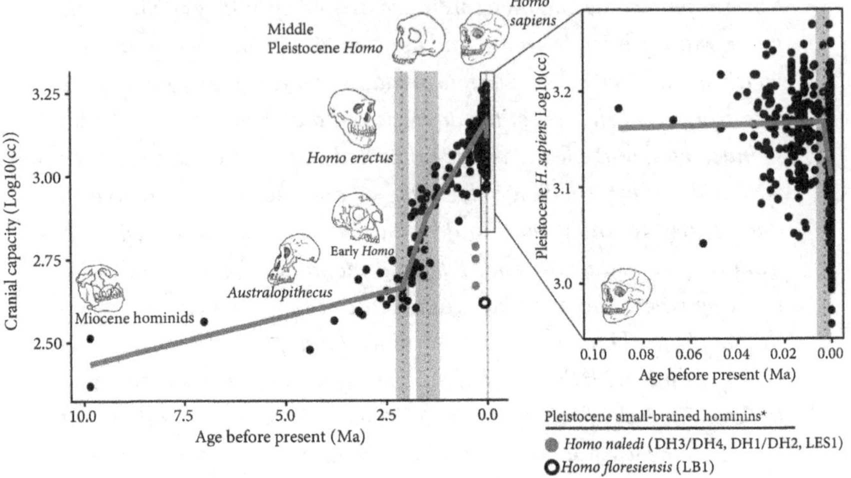



**Figure 5.1** Change in hominin cranial capacity over the past ten million years. (Source: Data from J. M. DeSilva et al., *Frontiers in Ecology and Evolution* 21 [2021]: 742639.) Note that there was a rapid increase in brain volume from *Australopithecus* to modern *Homo sapiens*. The right-hand panel shows cranial capacity over the last one hundred thousand years.



thing you should notice is that over the history of our ancestors, brains have more than tripled in size, and our skulls are more than three times the size of our closest relatives—the chimpanzees and bonobos, and our australopithecine ancestors.

In some types of organisms like plants, natural selection acts on the size of the offspring—in this case seeds—to optimize them for survival in the environments where they live. Orchids have seeds that are almost like dust, so they are easily dispersed. They have a symbiotic relationship with a fungus that enables them to germinate and grow into an adult plant. For plants that live in shady forests, like the avocado, seeds have to be large enough for the seedling to make leaves that can capture enough light for survival. The largest seeds are made by a palm called coco de mer—these coconuts are a foot long and can weigh forty pounds each. Their large size is an adaptation for floating in the ocean for dispersal and then producing roots that can grow deep enough to reach fresh water once they are washed up on a beach. The seeds of plants vary in size by several orders of magnitude—the evolution of seed size appears to be much less constrained than the offspring of vertebrates such as birds and mammals.

For birds, there are evolutionary constraints on egg size because of the limits of the size of their cloaca, which acts as the birth canal the same way the vagina does in mammals. The egg can only be so large, so when it hatches, the chicks of most bird species are much too immature to fly or obtain food on their own. Depending on the species, they require a relatively long period of care before they are strong enough to learn to fly and forage for food. In contrast, hatchlings of turtles, lizards, snakes, and alligators require little or no parental care—they are able to forage for their own food as soon as they hatch. Those that are born developmentally early require much more parental care before they can survive on their own. Animals differ in how mature they are when they are born or hatched; if they require a lot of parental care, then we say they are *altricial*, while newborn offspring able to survive on their own are *precocial*. We see variation within birds for how altricial their hatchlings are based on the environment they live in and the constraints of egg size relative to the size of the adult. In many species of birds, the physical constraints on egg size are more prevalent, so more parental care is required before hatchlings can survive on their own.

Among primates, most species have offspring that are developmentally advanced—they are precocial—compared to human babies. Newborns of chimps, gorillas, and monkeys have enough muscle coordination and strength to be able to hold on to the fur of their mothers and ride on their backs. Human babies, on the other hand, are helpless—they are altricial—and require years of care before they can possibly survive on their own. Why do you think our ancestors lost the ability to birth developmentally advanced infants? Was this adaptive? Something that improved their ability to survive in the savanna?

If you look at Figure 5.2, which shows a diagram of the pelvic bone structure during childbirth, you'll notice that the infant's head seems to barely fit through the pelvis. During childbirth, the pelvis stretches and the infant's head has to squeeze through the birth canal. An infant's bones are still flexible when they are born; the tight fit of the mother's pelvis is why babies often have elongated heads after delivery. It appears that there is a constraint on the head size of newborn infants, but there is also speculation that metabolic constraints of pregnant females may have contributed to earlier births.[3] On the other hand, in the months following birth, head size increases rapidly, which is more consistent with the idea that there is a constraint on head size at birth.[4] Either way, the important thing is that infants of our *Homo* ancestors were much more altricial than other primates and required several years of care before they could feed themselves. As

---

[3] H. M. Dunsworth, "Metabolic Hypothesis for Human Altriciality," *Proceedings of the National Academy of Sciences* 109 (2012): 15212–16.

[4] P. B. Gray and J. R. Garcia, *Evolution and Human Sexual Behavior* (Harvard University Press, 2013).

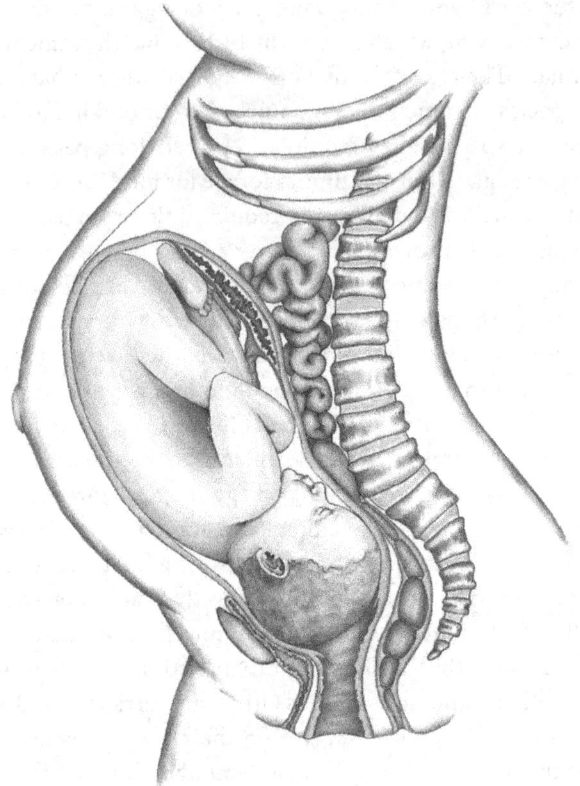

**Figure 5.2**  Human pelvic girdle during birthing.
Katherine Wonder (own work).

selection favored increased brain size in our ancestors, it forced our infants to be born developmentally earlier, and consequently the amount of care that infants required dramatically increased. This single change—earlier births resulting in more altricial infants—had multiple consequences, including major changes to the mating practices and social structures of the clans of our ancestors.

The difference in body size between males and females can tell us a lot about their mating habits. In gorillas, there is a large difference in the body size of males and females, with males being about 50 percent larger by weight than females. Such a large body size difference is usually associated with polygyny, where one male has control over a harem and mates with multiple females. This is something that Charles Darwin noticed about gorillas—where males are much larger than females—and he speculated that it had to do with the fact that they are polygynous. A large body size difference is known in other animals with polygynous mating, including some marine mammals such as sea elephants. Selection favors a larger body size of the male because he has to defend his harem from other males who are eager to replace him. Some early hominin species such as *Ardipithecus* and *Paranthropus* may have been polygynous,[5] but this idea is speculative because the biomarker used (comparison of finger lengths) is only loosely associated with polygyny in primates, and there are too few fossil samples available for these hominin species to accurately estimate male and female body sizes. The situation is similar for *Australopithecus* species like Lucy. Early studies indicated that there was a large difference in body size between males and females, but more recent analyses suggest that the difference may be much smaller. We only have evidence from a few individuals, and there may have been other selective forces that limited the size difference between males and females, so we may not know whether there was a large size difference until more specimens are found. As we will discuss later in this chapter, it's more likely that Lucy's species lived in multimale/multifemale groups like chimpanzees.

We know that monogamy is associated with small differences in body size between males and females. By the time of the appearance of *H. erectus*, we see that males and females are similar in size. In this species and in modern humans, on average males are only about 15 percent larger by weight than females. Based on the difference in body size between males and females, it's likely that our more recent ancestors such as *H. erectus* were monogamous, just like humans are today in most cultures around the world. While there are several types of monogamy, it generally means that two people cohabitate, maintain a sexual relationship, and share resources and parenting responsibilities. Depending on cultural practices, sex outside the monogamous relationship may be acceptable, and monogamous relationships may not be lifelong (*serial monogamy*). Serial monogamy may have been prevalent in our ancestors,[6] particularly because death during childbirth and due to accidents and disease were common. Why do you think monogamy was favored?

For hominin species with larger heads like *H. erectus* and *Homo sapiens*, the best explanation for the transition to monogamy is that their babies required more care; it took two people to raise infants and to ensure their survival. In polygynous species like gorillas and chimpanzees, only the female cares for her infant, and the amount of time it takes until infants are mature enough to care for themselves is much shorter. As infants required more parental care in our

---

[5] E. Nelson et al., "Digit Ratios Predict Polygyny in Early Apes, *Ardipithecus*, Neanderthals and Early Modern Humans but Not in *Australopithecus*," *Proceedings: Biological Sciences* 278 (2011): 1556–63.

[6] R. Schacht and K. Kramer, "Are We Monogamous? A Review of the Evolution of Pair-Bonding in Humans and Its Contemporary Variation Cross-Culturally," *Frontiers in Ecology and Evolution* 7 (2019). https://doi.org/10.3389/fevo.2019.00230

*Homo* ancestors, monogamy was favored. We know that infants of Lucy's species were probably more like chimpanzee infants and required only a few months of breastfeeding.[7] Consequently, it is unclear whether the small difference in body size between males and females in *A. afarensis* is actually indicative of monogamy; based on the amount of childcare required, it is more likely that they were polygynandrous, where both males and females have multiple mates, and lived in multimale/multifemale groups like chimpanzees.

Later hominins in the genus *Homo* had much larger heads, and participation in childcare by both parents improved the infant's chance of survival; there was selection on male behavior to be good providers and caregivers. Parental care by both parents became increasingly important as brain size increased because it took longer for offspring to become independent—infants required more and more care to ensure their survival. While it has been suggested that the finger-length biomarker indicates polygyny in some early species of *Homo* including Neanderthals,[8] this is unlikely given the amount of parental care that was required to ensure the survival of infants. We can conclude that monogamy in modern humans is a product of evolution by natural selection, because it improved the chances of survival of our ancestors' offspring.

In evolutionary biology, we measure the success of individuals as their ability to successfully raise offspring, which we generally refer to as their *fitness*. In an evolutionary context, fitness is not a measure of physical strength but reflects all the characteristics of an individual that improve their survival and the number of offspring they produce. Fitness is a measure of the ability of individuals to pass their genes on to their children. As head size increased, hominin infants required more care, so selection favored caregiving by both parents. But monogamy on its own was not enough to ensure the survival of infants.

Under some circumstances, cooperation among individuals of a species is favored by selection. For example, many species of ground-dwelling animals have individuals that serve as sentries to watch for predators. In species such as ground squirrels, prairie dogs, and meerkats, older males will stand guard and give a warning call if they see a predator. They ensure the survival of other members of the colony even as they increase the risk to their own lives. Evolution tells us that they should watch out for their own survival, so why would they risk their lives to ensure the survival of others?

We generally view selection as favoring behaviors that improve the fitness of the individual, and if we only consider the individual, it doesn't make sense that

---

[7] T. Tacail et al., "Calcium Isotopic Patterns in Enamel Reflect Different Nursing Behaviors among South African Early Hominins," *Science Advances* 5 (2019): eaax3250.

[8] E. Nelson et al., "Digit Ratios Predict Polygyny in Early Apes, *Ardipithecus*, Neanderthals and Early Modern Humans but Not in *Australopithecus*," *Proceedings: Biological Sciences* 278 (2011): 1556–63.

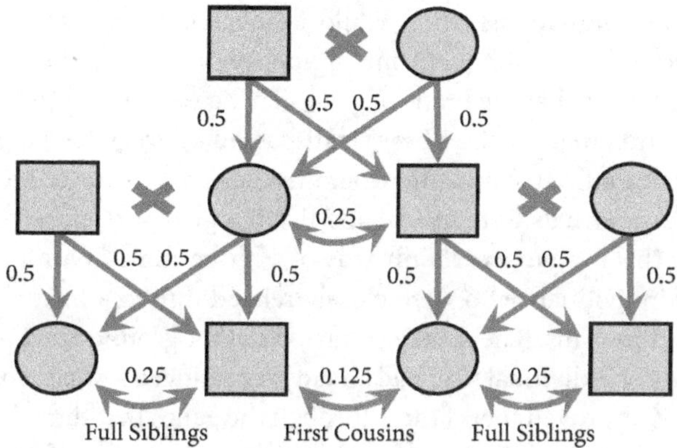


**Figure 5.3**  Levels of relatedness (gene sharing) within an extended family. M.B. Cruzan (own work).

selection would favor risky behaviors that do not increase individual fitness. But there is something else to consider; it's likely that some of the squirrels in the colony are the sentry's offspring. From an evolutionary point of view, should a squirrel risk being eaten by a hawk if they can save their own offspring? Sure, because if some of the other squirrels were their children, it makes sense that the sentry squirrel would warn them about the predator even if it reduced his chances of survival. But how do the sentry's children differ from a random squirrel? The answer is that the sentry shares a lot of genes with their children. In fact, they are related by one-half, meaning that each offspring has half of the parent's genes. Since they share a large proportion of their genes with their offspring, doing something to ensure they survive increases fitness—even if it means they might be the one who gets eaten. This idea is called *inclusive fitness*, or sometimes *kin selection*, because it focuses on genes that are shared among related individuals. Remember that the whole point of evolution—the way that individuals increase their fitness—is to get copies of their genes into the next generation. It doesn't matter if they do that through their own reproduction or by making sure closely related individuals survive and reproduce. The result is the same.

Now let's think about this larger family tree, which includes parents, offspring, and cousins (Figure 5.3). As we said before, the relatedness between parents and offspring is one-half, and it's also one-half between children with the same parents. For children sharing one parent, the relatedness is one-quarter, and for pairs of cousins it's one-eighth. Notice that there is a lot of gene sharing among individuals of the same family, and even for extended families.

A colony of squirrels would likely have been in the same meadow for a long time, so over many generations, as more inbreeding occurs due to mating between related individuals, they would all become more related; over generations, there might be frequent mating between cousins and even between siblings. This means that the levels of gene sharing would be much higher in an inbred population than the numbers we discussed above, which are for a population that is not inbred. The same would have been true for our ancestors; we know that Neanderthals were inbred and lived in groups of closely related individuals,[9] and this is almost certainly true of all hominins.[10] Our ancestors lived in groups of individuals who were closely related due to a history of inbreeding; they lived in clans. The sentry behavior of that ground squirrel—which is typically an older individual who had already reproduced—was favored by selection because it improved their fitness through the genes that they shared with all individuals in the colony. In the same way, cooperation and self-sacrifice for the benefit of others would have been favored in our human ancestors because they increased their inclusive fitness.

The degree to which cooperation was favored in recent hominins in the genus *Homo* compared to our earlier hominin ancestors had much to do with how much childcare was required to ensure the survival of offspring. Through analysis of the chemical composition of teeth, we know that infants of hominin species required different amounts of care.[11] It turns out that breast milk contains specific calcium isotopes that are deposited in tooth enamel, so the amount of these isotopes present indicates how long infants were breastfed before weaning. Studies of the isotope composition of teeth in species like *Australopithecus*, including Lucy's species, indicate that infants were weaned after only a few months of breastfeeding. This suggests that Lucy's species was more similar to chimpanzees than humans, and much less childcare was required before their offspring could feed themselves. Even though males and females of Lucy's species may have had similar body sizes, it's unlikely that monogamy would have been favored. By the time we get to hominins in the genus *Homo*, the time to weaning is much longer, between two and four years. And it's the same for our most recent ancestors.

Think about it—from a few months to four years of care. Do you think that just having two parents would be enough to ensure the survival of infants in harsh environments such as the savanna? Because of historical inbreeding, members of the clan share large numbers of genes with each other. Each member can then increase their fitness by helping out with gathering and preparing food and caring for children and ultimately improve their chances of survival. Over time,

---

[9] L. Skov et al., "Genetic Insights into the Social Organization of Neanderthals," *Nature* 610 (2022): 519–25.

[10] M. Grove et al., "Fission-Fusion and the Evolution of Hominin Social Systems," *Journal of Human Evolution* 62 (2012): 191–200.

[11] T. Tacail et al., "Calcium Isotopic Patterns in Enamel Reflect Different Nursing Behaviors among South African Early Hominins," *Science Advances* 5 (2019): eaax3250.

those clans that had higher levels of cooperation for food gathering and caring for infants had faster growth rates and eventually split into more clans. The lineages with low levels of cooperation died out because their infants had lower survival rates. As brain size increased in our ancestors, cooperation was favored.

There is another idea related to inclusive fitness and cooperation in inbred populations that may have been important for the survival of our hominin ancestors. In most species of mammals, females remain reproductive throughout their entire lives. The two exceptions to this are toothed whales, like orcas, and humans. In both of these species, females go through menopause and spend significant portions of their lives without the ability to reproduce. Males of the species, on the other hand, are able to produce viable sperm their entire lives. The *grandmother hypothesis* is the idea that the presence of postmenopausal females increases the chances of survival of the young, and in doing so, it also increases the fitness of the grandmothers—their inclusive fitness. In orcas, the older females help the younger individuals by showing them where and how to obtain food. What do you think? Does this idea sound like a reasonable explanation for menopause in our hominin ancestors?

Unfortunately, we know from fossil evidence that our early ancestors rarely lived past the age of forty,[12] which is earlier than the average age for the onset of menopause. Assuming the onset of menopause would have been similar in our ancestors in the genus *Homo*, it's unlikely that there were many postmenopausal grandmothers around to participate in childcare. It's surprising that other authors who have promoted the grandmother hypothesis[13] have not recognized that very few postmenopausal women would have been present in the clans of our *Homo* ancestors. The grandmother hypothesis can be may be more relevant in more recent societies where women usually live past menopause, but it is certainly not universally true, because not all grandmothers participate in childcare. In our early ancestors it would have been rare for women to live past menopause, so it is unlikely that the grandmother hypothesis is an explanation for the evolutionary origin of menopause in our hominin ancestors. We know that our closest living relatives—the chimpanzees and bonobos—do have menopause in captivity,[14] but it is probably uncommon in the wild because lifespan is shorter. Much more recently, it's possible that females of *H. sapiens* started living longer, and care by nonreproductive grandmothers was favored because it improved the survival of grandchildren.

---

[12] J. N. Kotre and E. Hall, *Seasons of Life: The Dramatic Journey from Birth to Death* (University of Michigan Press, 1997).

[13] J. Diamond, *Why Is Sex Fun?* (Basic Books, 1997).

[14] M. L. Walker and J. G. Herndon, "Menopause in Nonhuman Primates?," *Biology of Reproduction* 79 (2008): 398–406.

So why do human women live past menopause while other apes do not? The answer probably has to do with two things: more frequent ovulation and longer lifespans in our recent history. We can understand this better if we compare ourselves to the other great apes. Females are born with a certain number of eggs, and menopause commences once those eggs are depleted or nearly so. Other great ape females ovulate only a few times per year, while human females ovulate about once a month after puberty. Even though human females start out with many more eggs, they deplete their egg supply more quickly. If our ancestors had routinely lived past the age of forty, then we could predict that modern human females were born with many more eggs to ensure their reproductive ability throughout their expected lifespans. The higher frequency of ovulation meant that our early ancestors had a greater capacity for rapid population growth. This may have resulted from very low rates of survival at some point in our history so that a higher reproductive capacity was necessary to ensure the persistence of our hominin ancestors. There was no selection to increase egg number in our ancestors because women only rarely lived past menopause. Much later, survival improved as our ancestors gained more advanced skills and technologies. Since their reproductive capacity remained high, their population growth rates increased, resulting in their dominance in many habitats as they spread across the globe. Just like chimpanzees held in captivity, improved technologies increased rates of survival, which meant longer lifespans in our recent human ancestors. As more women have lived into their menopausal years, they may have improved the survival of their grandchildren.

The grandmother hypothesis may apply to our very recent ancestors, and this same idea may explain something else that is unique to humans. To understand the selection pressures leading to the evolution of different sexual and gender identities, we first need to talk about naked mole rats. These animals live in underground colonies in large systems of connected tunnels in the extremely dry and harsh environment of the East African grasslands. Only a few individuals participate in reproduction, and most members of the colony are workers who gather food—they act as helpers.

In the case of the naked mole rat, it is known that the helper role for most individuals is favored to ensure survival of the colony. If all individuals reproduced, there would be many more offspring, but infant mortality would be high and the colony would probably not survive, so natural selection has favored having members that do not participate in reproduction. Since colonies are inbred, this behavior is favored through increases in inclusive fitness. We see the same type of cooperation in colonial nesting birds, such as the acorn and Lewis's woodpeckers, bushtits, azure-winged magpies, and ground hornbills; in several mammals, including nine different genera of rodents, meerkats, and wolves; and in two primates (marmosets and tamarins). The two woodpecker species live in colonies in western North America; like other cooperative breeding species, younger individuals do not reproduce but act as *helpers at the nest*.[15] This behavior is favored because the helpers are typically siblings to the nestlings, so there is an inclusive fitness advantage for them. They can't reproduce on their own because there are a limited number of nest holes available. For both the mole rats and these woodpeckers, selection has favored including nonreproductive individuals to act as helpers in resource-limited environments and, consequently, to ensure the reproduction and survival of the entire colony.

One of the puzzles of human evolution is the high frequency of homosexuality in our species; the rate of people who identify as lesbian or gay is between 3 and 16 percent, and this is fairly consistent for populations around the world.[16] It's important to recognize that the high degree of variation in reported frequencies among cultures may have more to do with the acceptance of homosexuality; the actual rate is probably much higher than the lower range that these numbers suggest. More importantly, humans are one of only two species that have substantial numbers of individuals who are *exclusively homosexual* and do not participate in reproduction.[17] Exclusive homosexuality differs from homosexual behavior because these individuals never willingly engage in heterosexual sex. The only other species that has exclusively homosexual individuals is the domesticated sheep, and there are no known examples of exclusive homosexuality in wild species, including the Asiatic mouflon, the ancestor of domestic sheep.[18] In sheep, the occurrence of exclusive homosexuality is probably an accidental outcome of reduced population size during domestication, and it has been maintained in sheep herds because they are buffered from the effects of natural selection. Other primates, especially bonobos, have homosexual behaviors, but all individuals of these species are actually bisexual and engage in heterosexual copulation as well. While we do not completely understand the genetic determination of homosexuality in humans, it is clear that exclusive homosexuality is unique and intrinsic to our species. How do we explain its high frequency?

The grandmother hypothesis may not provide much of an explanation for menopause in women, but it's likely that the harsh environments where our early *Homo* ancestors lived would have favored including homosexual members in the

[15] S. T. Emlen, "Evolution of Cooperative Breeding in Birds and Mammals," in *Behavioural Ecology: An Evolutionary Approach*, 3rd ed., ed. J. R. Krebs and N. B. Davies (Blackwell Scientific Publications, 1991), 301–37.

[16] *Between Men: HIV/STI Prevention for Men Who Have Sex with Men*, OCLC 896761012 (International HIV/AIDS Alliance, 2003).

[17] A. Poiani and A. F. Dixson, *Animal Homosexuality: A Biosocial Perspective* (Cambridge University Press, 2010).

[18] B. Bagemihl, *Biological Exuberance: Animal Homosexuality and Natural Diversity* (St. Martin's Publishing Group, 2000).

clan; these individuals acted like "grandmothers" because they were helpers, and they did not participate in reproduction. Like the naked mole rat workers and the woodpecker siblings, they could help ensure the survival of the clan by hunting for animals, gathering roots and grain, and caring for the young. Through their efforts to ensure the survival and reproduction of others, homosexuals would have increased their fitness through their relatedness with other members of the clan. This is similar to the cooperation we discussed earlier; those clans that embraced their homosexual members would have been more successful, so this trait would have spread.

An obvious difference between the helpers in the clans of our ancestors and nonreproductive helpers in other species such as mole rats is their sexual relationships with members of the same sex. One might ask why evolution favored this particular solution to ensure that clans of our *Homo* ancestors had adequate numbers of nonreproductive helpers. An important difference between humans and most other animals is our constant sex drive after puberty. Other species have sex drive only in particular seasons or under specific conditions; at all other times, their drive is "turned off." Having our drive "turned on" all the time appears to be an intrinsic characteristic of our species that may have arisen when food was available year-round and when there was selection for a fast population growth rate. It has been speculated that constant sexual desire and frequent copulation were important for pair bonding to maintain stable monogamous relationships as our offspring required increasing amounts of parental care. In any case, since sex drive in other species is periodic, it may have been easier for selection to repress it in helpers. Humans, on the other hand, have a constant sex drive, so the easiest evolutionary path to generate nonreproductive helpers would have been to redirect sexual desires toward members of the same sex. Hence, exclusive homosexuality in our *Homo* ancestors arose as our offspring became increasingly altricial and more help was needed to ensure their survival and the survival of the entire clan.

Some authors do not acknowledge the high frequency of exclusive homosexuality in humans.[19] They have erroneously concluded that homosexuality was maladaptive and that it could only persist because homosexuals also had heterosexual relationships. This view is obviously incorrect because it does not explain exclusive homosexuality or consider the potential for inclusive fitness in our ancestors that favored homosexual individuals acting as nonreproductive helpers. Without these considerations, we cannot understand how homosexuality would persist in human populations unless all homosexuals were actually bisexual.

[19] R. Baker, *Sperm Wars* (Basic Books, 2020).

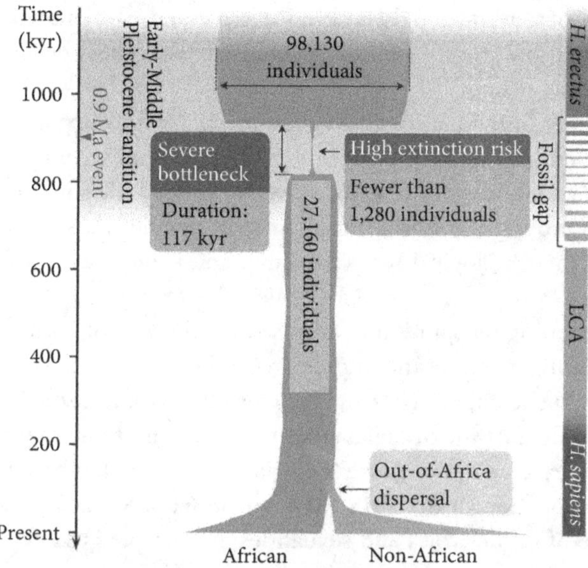


**Figure 5.4** Demographic history of human ancestors over the past million years based on variation in the nuclear genomes of modern humans. The width of the diagram indicates estimated historical population size, showing a severe population bottleneck around nine hundred thousand years ago in our *Homo erectus* ancestors.

Reproduced with permission from Wangjie Hu et al., *Science* 381 (2023): 979–84, https://doi.org/10.1126/science.abq7487.

The presence of homosexuals in cultures around the world indicates that this characteristic arose early in our *Homo* ancestors, prior to the major migrations out of eastern Africa around 70,000 years ago that led to the colonization of the remainder of Africa and all other continents except Antarctica. Detailed analyses of variations in human genomes indicates that there was a severe reduction in population size in our ancestral lineage around 900,000 years ago that lasted 100,000 years (Figure 5.4).[20] During this time, eastern Africa was subjected to severe drought associated with glaciation in northern latitudes. As you can see in Figure 5.4, our ancestral lineage was reduced to just over 1,000 individuals and nearly went extinct; this event could explain the origin of exclusive homosexuality and other modes of cooperative behaviors in our *H. erectus* ancestors. The inclusion of exclusively homosexual individuals in clans would have increased

---

[20] Wangjie Hu et al., "Genomic Inference of a Severe Human Bottleneck during the Early to Middle Pleistocene Transition," *Science* 381 (2023): 979–84.

their chances of survival and their ultimate success; these clans would have been more likely to grow and spread into new territories as they repeatedly split into multiple daughter clans.

The idea that individuals with a range of gender and sexual identities were important members of ancestral clans prior to the major migrations starting 70,000 years ago is supported by the observation that they are present in modern societies around the world. Acceptance of people whose gender identities don't match their biological sex is growing, and some societies recognize and respect members with diverse gender roles. For example, in Samoa, there has been long-standing recognition of *fa'afafine* (in the way of a woman) as a third gender of effeminate males and *fa'afatama* (in the way of a man) as a fourth gender of masculine females.[21] Note that gender identity and sexuality are separate traits, so not all fa'afafine or fa'afatama individuals are homosexual. A fa'afafine soccer star is featured in the movie *Next Goal Wins*, which is based on American Samoa's participation in the 2014 FIFA World Cup. As in Samoa, individuals with a range of gender roles and sexualities are accepted as equal members in numerous societies around the world.

It's unclear why homosexuality and equal participation of both genders in all roles became taboo in many cultures in our recent history. The idea of scripted roles for men and women—and the false idea of female inferiority—was taught in Christianity, Judaism, Islam, and many other religions that promoted patriarchal societies, but the assertion of male superiority is not a consistent feature across Eastern religions.[22] In Western cultures and some Eastern cultures, men used, and continue to use, religious scripture to justify their false claims of superior intelligence and their superior status in society. Similarly, religious scripture has been used to rationalize the rejection, persecution, and sometimes even murder of homosexuals because they were defined as "unnatural." But as we have learned here, nothing could be further from the truth. Our natural social structure can be modeled from the clans of our ancestors, where men and women participated in all roles and where the presence of homosexual individuals was important for survival.

Fortunately, many societies around the world are returning to these natural conditions. We are seeing acceptance of a wide range of male and female roles and a wide range of both gender and sexual identities. Not surprisingly, entrenched religious conservatives are pushing back and are claiming that the diversity of gender identities is somehow an insidious byproduct of progressive idealism. But the opposite is true. Homosexuals as well as individuals with

[21] P. L. Vasey and D. P. VanderLaan, "Fa'afafine," in *Encyclopedia of Evolutionary Psychological Science*, ed. T. K. Shackelford and V. A. Weekes-Shackelford (Springer, 2021). pp. 2875–76.
[22] L. Woodhead, *Religions in the Modern World: Traditions and Transformations* (Psychology Press, 2002).

diverse sexual and gender identities have always been present; they are a natural and important facet of humanity. It's only in an increasing atmosphere of acceptance that these individuals are becoming more comfortable and consequently more visible. The increasing acceptance of men and women performing all roles as well as LGBTQ+ individuals in our society is a return to a normal state, one that was favored by natural selection in the clans of our ancestors.

We can't know for sure how many of our human characteristics have evolved, so it's important to understand how we can draw conclusions about things that we cannot test using experiments. In evolutionary science, we develop explanations for traits and behaviors by following the paths that are best supported by observations and best guided by evolutionary logic. In the case of cooperation and exclusive homosexuality, our conclusions are supported by other examples in nature—like the mole rats and woodpeckers—that demonstrate that there are evolutionary advantages to having nonreproductive helpers when resources are limited. We are guided by evolutionary logic; because the group is inbred, the individual helpers—grandmothers and homosexuals—gain fitness through inclusive fitness due to gene sharing across the clan. It follows that clans that included helpers had higher rates of survival and growth. Together, these observations make a strong argument to support the hypothesis that cooperation and nonreproductive helpers were favored by natural selection and were important for the survival of our ancestors. This would have been most important during periods of severe drought when cooperation would have been critical for survival of the clan. As mentioned above, such an episode occurred starting 900,000 years ago and lasted more than 100,000 years, resulting in the near extinction of our ancestral lineage.[23] The advantage of having homosexual individuals as clan members must have continued as they were favored by selection during migrations that colonized continents around the world. These are the facts as we know them, and they are the best explanations we have for the origins of cooperation, monogamy, and homosexuality in humans.

Now, contrast the clans of our ancestors with modern society. Today, many of us live in cities and towns populated by mixtures of people from many different places. Especially in industrialized societies, it is no longer true that your neighbor or the people you interact with every day are your close relatives, so our acts of kindness, and even self-sacrifice for others, appear to be altruistic because there is no inclusive fitness benefit. The instinct for helping others— even risking your own life to save someone—is so deeply engrained in our psyche by thousands of generations of selection that we view unselfish acts as normal.

[23] Wangjie Hu et al., "Genomic Inference of a Severe Human Bottleneck during the Early to Middle Pleistocene Transition," *Science* 381 (2023): 979–84.

Since 1904, the Carnegie Hero Fund has awarded medals to more than 9,500 people for putting themselves at risk to save the lives of others. A recent study of these heroes showed that amygdala region of their brains was often slightly larger than average,[24] suggesting that these individuals have greater capacity for empathy. Almost all of these heroes say that they acted without thinking and that it was an instantaneous decision—like an instinct.

But not all unselfish acts are so dramatic. Charity organizations rely on donations from the public, and they are able to operate because, thanks to our ancestors, we have been ingrained with an instinct to help others. We are the only species that exhibits behaviors that are seemingly altruistic—to help others with no material expectation of a benefit, except that it makes us feel good. The most logical explanation for this behavior is that it originated in our hominin ancestors in the dry woodlands and savannas of Africa. As our ancestors' brains increased in size, their offspring required more care, and cooperation among members of the clan was favored through the evolutionary advantages of inclusive fitness. But today we live among strangers, and our instinct to help others looks altruistic.

But don't you think that our intelligence has something to do with our helping others? That we are smart enough to understand that giving donations to charities is good for society? Of course, our instincts to be generous have developed further; they are codified in our ethical and religious belief systems and in the behavior we expect of each other as members of modern societies. All we're saying here is that the origin of altruism in humans comes from our history, from the many thousands of generations when cooperation was critical for the survival of the clans of our hominin ancestors.

A competing hypothesis for increases in cooperation in our human ancestors is the idea that there was "self-selection" for increased "tameness" similar to changes we see in the domestication of animals such as cats and dogs compared to their wild ancestors[25]. The idea of *Self-Domestication* of humans was first proposed around 1800 by Johann Friedrich Blumenbach[26], who is best known for his outdated and false ideas about the classification and origin of human races. The advocates of self-domestication point to the similarities in domestic animals and humans including flattened faces (shorter snouts) and smaller teeth (especially canines). Some have also claimed that self-domestication is supported by the identification of genetic differences between humans and other

---

[24] A.A. Marsh et al., "Neural and Cognitive Characteristics of Extraordinary Altruists," *Proceedings of the National Academy of Sciences* 111 (2014): 15036–41.

[25] Wrangham, R. W., "Hypotheses for the evolution of reduced reactive aggression in the context of human self-domestication," *Frontiers in Psychology* 10 (2019): 1914.

[26] Wrangham, R., "The Goodness Paradox: The Strange Relationship Between Virtue and Violence in Human Evolution." (Pantheon Books 2019).

hominins controlling embryonic development that affect facial features in modern humans[27]. But finding genes responsible for facial features is not the same identifying the cause of their appearance in our ancestors; this discovery does not support the self-domestication hypothesis. Darwin also recognized the fallacy of self-domestication as originally proposed by Blumenbach and rejected it simply because there was no selective force available to explain it. In other words, there was no fitness advantage for increased tameness or particular facial features in our *Homo* ancestors. Furthermore, tameness—reduced aggression—is not necessary for high levels of cooperation. For example, inbred groups of wild animals, including the ancestor of domestic dogs, the gray wolf, display a wide range of cooperative behaviors[28]. As previously discussed, observed changes in the morphology of skulls, jaws, and teeth in our ancestors is consistent selection due to changes in diet and increase in brain size after consumption of cooked food became common. We can also attribute cooperation in the inbred clans of our *Homo* ancestors to increases in inclusive fitness as much greater amounts of childcare were necessary for infant survival as brain size increased.

By now you should understand the puzzling statement I made at the beginning of this chapter: because of increases in brain size, our infants were born earlier in their development and required more and more care. Inbreeding within clans increased the effectiveness of selection for increased cooperation through inclusive fitness. Consequently, monogamy, cooperation, and the inclusion of homosexual individuals as helpers were favored to ensure the survival of the young and the entire clan.

*Summary*—In this chapter we learned that australopithecine hominins such as Lucy's species had much smaller brains, so their offspring were much more precocial and more similar to chimpanzee infants. Consequently, females nursed their infants for only a few months and males probably did not participate in childcare. As brain size increased in early species of *Homo*, infants were born developmentally earlier; they became increasingly altricial and required years of care instead of months. Consequently, monogamy and increased participation of both partners in childcare were favored.

We reviewed the concept of inclusive fitness; individuals can increase their own fitness through behaviors that increase the fitness of related individuals. Because the clans of our ancestors were inbred, there were high levels of gene sharing among individuals, and consequently, cooperation was favored.

[27] Zanella, M. et al., "Dosage analysis of the 7q11.23 Williams region identifies BAZ1B as a major human gene patterning the modern human face and underlying self-domestication," *Science Advances* 5 (2019): eaaw7908.

[28] Cassidy, K. and McIntyre, R.T. May, "Do gray wolves (*Canis lupus*) support pack mates during aggressive inter-pack interactions?" *Animal Cognition.* 19 (2016): 939–47.

The instinct for cooperation and helping others was critical for the survival of our ancestors, but in modern societies these behaviors appear altruistic.

The grandmother hypothesis suggests that menopause in toothed whales and humans was favored through increased inclusive fitness for postreproductive females who helped care for their grandchildren. We concluded that this idea is flawed for humans because it does not take into account the fact that our hominin ancestors rarely lived past the age of forty. Consequently, menopausal females were rare and did not occur at high enough frequencies for selection to be considered the cause of menopause in our *Homo* ancestors. It's more likely that menopause is a consequence of the lack of selection to maintain reproductive ability for females of our ancestors past the age of 40.

We learned that humans are only one of two species that include individuals who are exclusively homosexual and that homosexuals are members of human societies worldwide. By comparing our early ancestors to other species that have nonreproductive helpers in systems of cooperative breeding, we realized that inclusion of homosexuals would have increased the survival of clans living in harsh environments. As head size increased in our ancestors, infants required increasing care and monogamy was favored. Inbreeding and inclusive fitness favored cooperation among clan members and inclusion of homosexual individuals as nonreproductive helpers. Clans that included homosexual members would have had higher survival of offspring and consequently would have spread to displace clans with low levels of cooperation and that did not include homosexual individuals. It's likely that a 100,000-year period of severe drought that occurred 900,000 years ago and caused the near extinction of our ancestral lineage served as a catalyst to establish our instincts for cooperation, as well as the inclusion of exclusively homosexual individuals in our ancestral clans.

# 6

# It takes a clan

*Over the days that followed the birth of her baby, Launua learned about how to care for him. There were older men and women who helped her with breastfeeding and cleaning the baby. They showed her how to make a swaddle, so she could have her hands free to help pre-pare food. It had never occurred to her until now, but the two men who helped with childcare and cooking shared the same sleeping place and did not look at her the same as the other men. She also noticed that a few of the women had no interest in men. They preferred to go hunting instead of gathering or cooking, and her partner said that one of them was the best hunter in the clan.*

*Launua was glad that everyone worked together, and things seemed to go well for many days. But then, the rains did not come. The plants dried up and the hunters said they could not find animals—there was no meat for many days. There was very little food and Launua could feel the hunger in her stomach. She felt her body becoming thinner, and she did not have enough milk for the baby. Everyone seemed weak and hungry, and one of the other babies died. Launua hugged her*

*Looking Down the Tree*. Mitchell B. Cruzan, Oxford University Press. © Oxford University Press (2025).
DOI: 10.1093/9780197805190.003.0006

*infant close and tried to squeeze milk from her breast to feed him. The hunters still went out looking for prey, and one day they came back with an animal. It was thin with little fat, but it was enough for them to get by until another animal was killed. Then the rains came, and everything seemed fine again. Launua had never realized how important her clan was—her baby would have never survived without their help. She thought about how well everyone seemed to work together to provide food and to make sure all the members of the clan had their share. She was happy now that life was good and grateful to be a part of her clan.*

If we look at pictures of great apes other than humans, it is often difficult to tell the difference between males and females because all great ape females except humans are flat-chested. While the primary purpose of breasts is lactation and the feeding of offspring, the mammary glands, which produce milk, are only a small fraction of the mass in human breasts, which are mostly made up of fatty (*adipose*) tissue (Figure 6.1). The ability of a mother to produce milk for breastfeeding has only to do with her mammary glands and is not associated with breast size. So, we have to ask the question: why do human females have large breasts compared to other primates?

A number of ideas have been proposed for why human females have enlarged breasts and other primates don't. Some of the earliest writings used tactful

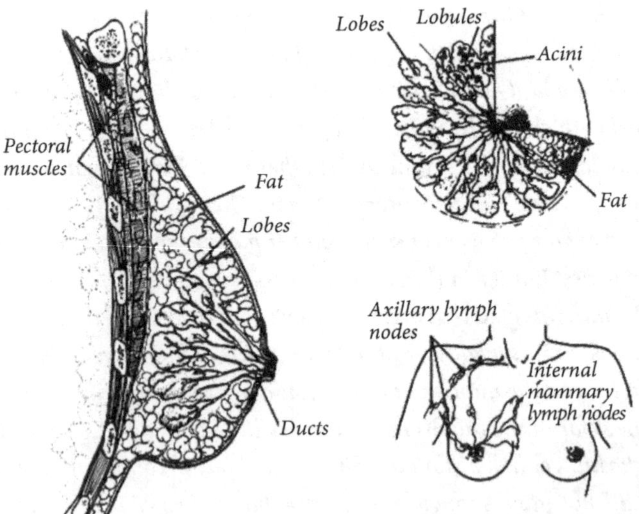


**Figure 6.1** Human female breast anatomy.

language to imply that men prefer women with large breasts, so clearly that is why they evolved. But this explanation is unsatisfactory for several reasons. First, it does not explain the wide variation in the sizes of breasts among women; if it were universally true that breast size was due to male preferences, wouldn't all women have very large breasts? We know that strong mate choice in other animals results in consistency of traits that are important for attracting mates. If males consistently preferred large breasts, then breast size would be consistently large. Second, the fitness of male hominin ancestors probably had more to do with how many females they mated with, so they wouldn't necessarily be choosy, but that's something we will discuss in a later chapter. Third, enlarged breasts are energetically expensive and unique to humans. It is much more likely that breasts in modern humans are the product of natural selection in our ancestors, but how?

The idea that breasts are objects of sexual interest is only true in cultures where women routinely cover their breasts. It's a little shocking to realize that so much of Western culture and the design of women's clothes is focused on breasts and buttocks. However, in many cultures, heterosexual men and lesbians don't view breasts as objects of sexual fascination simply because they see them all the time. In those cultures, breasts are "for the baby" and other parts of women's bodies are eroticized—whatever parts they cover when in public remain mysterious, and consequently they are the focus of sexual desires. In some cultures where women routinely wear less, it might be only the buttocks that pose sexual intrigue; in other cultures where women cover their heads and faces, almost all parts of the female body are considered erotic.

One idea on female breast size comes from the observations of an evolutionary biologist during the sexual revolution of the 1960s. This was a time when young people were generally rebelling against the conventions of the previous decades; men were growing their hair long and women were burning their bras. As part of this revolution, women were returning to breastfeeding instead of using the baby formula that had been sold to previous generations as being "technologically superior," and men were becoming more involved in childcare. One biologist wrote an account of watching his child while they were breastfeeding. He realized that the baby could hardly breathe because their nose was pressed against the breast as they suckled. He suggested that breast size and shape was a response to selection to move the nipple away from the mother's chest so the baby could breathe as they fed. In fact, the shape of most women's breasts actually interferes with an infant's ability to breathe while suckling.[1] This is not a problem when chimpanzee babies breastfeed; as you can see in

[1] E. Goldfield et al., "Coordination of Sucking, Swallowing, and Breathing and Oxygen Saturation during Early Infant Breast-Feeding and Bottle-Feeding," *Pediatric Research* 60 (2006): 450–55.

**Figure 6.2** Nursing human, macaque, and chimpanzee infants. Katherine Wonder (own work).

Figure 6.2 (bottom), their nostrils are so close to their eyes that they don't have a problem suckling and breathing at the same time.

The idea that breast shape may have been an adaptation to improve breathing in suckling infants does not explain why breasts are mostly fatty tissue and therefore energetically expensive. Think about how professional athletes in sports that require a lot of running, like soccer, marathon, and other track events, almost always seem to have an "athletic build" with small breasts. What a burden large breasts must have been for women thousands of years ago, working, hunting, gathering food, and running from predators—and all of that with no bra. We can think of a lot of reasons selection would have limited breast size as much as possible. So why are they sometimes so big?

Again, it's informative to compare ourselves to similar species. Like humans, some species of macaques have large noses that extend down their faces so the nostrils are close to their lips. As you can see in Figure 6.2 (middle), macaques don't have bulbous breasts; instead, they simply have elongated nipples that allow the baby to breathe while suckling. Consequently, we can conclude that large breasts have nothing to do with breathing at all. From what you have read in previous chapters, it should be pretty obvious: large breasts have everything to do with the transition to having babies earlier during development and having to breastfeed for years rather than months–especially during times when food supplies were unreliable.

Human breasts are energetically expensive to construct, and they can interfere with the ability of women to work and run quickly; just imagine what it would have been like for our ancestors trying to escape from a predator. Since they are costly, we can assume that breasts were favored by selection. But before we continue our discussion on the evolutionary origin of breasts and other aspects of the female body, let's talk about how evolution is not wasteful. When something is costly but not useful—when it does not increase fitness—selection will favor its reduction.

In many species we can see evidence of traits that are no longer useful and have been mostly or completely lost, and we call these *vestigial traits*. The cave salamander lives in complete darkness. It has no need for eyes, so they have been slowly reduced over generations to the point that now these salamanders have lost any evidence of eyes. This was probably due more to the absence of selection to maintain functional eyes rather than favoring individuals that had smaller and smaller eyes. Mutations that decreased eye size and functionality were free to increase in frequency in cave populations by random chance until the eyes of these salamanders were completely absent. The resources used to construct eyes could then be used for other purposes that increased fitness. We see similar vestigial traits in marine mammals such as whales. If you look at a whale skeleton, you will see where there were two small oblong bones just below the spine, close to the whale's tail. That's all that is left of the hips. Whales have no need for hind legs, so by mutation, random chance, and possibly selection, they were reduced until they were completely absent along with the reduction of the hip bones.

Naturally, it's always possible that a trait we call "vestigial" could have a purpose that we don't understand. But in the case of whales and other marine mammals, there are several lines of evidence that support the idea that these bones are vestigial hips. First, they are no longer connected to the spine, so they could not be important for swimming. Second, their position corresponds to the location we would expect to find hip bones in the whale's ancestors. And third, we have a fossil record of whale ancestors that could swim like whales but that still had hind legs. It's interesting to know that occasionally whales are born with vestigial hind limbs. This is called an *atavism* and it occurs in humans in rare cases where babies are born with vestigial tails. Atavistic traits such as legs on whales and tails on humans are not fully developed, and unlike the ancestral versions, they lack proper bone structure and muscularity so they are nonfunctional.

The leftover structures, such as the reduced eyes in the cave salamander and the hips in the whale, are vestigial and reflect structures that were fully functional in ancestors. The lesson here is that any structure that does not contribute to fitness is reduced and eventually lost as selection does not favor the maintenance of these traits; this allows for the allocation of resources to other functions that increase fitness.

The conclusion we can make is that the evolutionary function of human breasts has more to do with the large amount of fat they contain. In addition to the large amount of fat in breasts, women have a layer of fat just below the skin that is twice as thick as it is in men. They also have much thicker fat deposits in their thighs and buttocks. If we combine these observations with the fact that human infants require a long period of breastfeeding and care, and that our ancestors lived in harsh environments where food supplies were unreliable, we can see that these characteristics are connected. As brain size increased and infants became increasingly altricial, the large fat deposits in breastfeeding women would have been important for infant survival in our hominin ancestors when food was not available[2]—perhaps due to drought or harsh winters. We can conclude that breasts and other fat deposits in women represent adaptations to ensure the survival of offspring and are not just an out-come of partner preferences or a mechanism to allow babies to breathe while suckling.

The importance of fat deposits as food reserves for infant survival in early species of *Homo* is supported by the observation that newborn babies of modern humans have one of the highest levels of birthweight body fat among mammals.[3] Most mammals, including chimpanzees, have 2 to 3 percent of their birthweight as fat, while well-nourished human babies have around 15 percent of their birthweight as fat. The increasingly large brains of our *Homo* ancestors were energetically expensive to grow and maintain, so it makes sense that fat reserves in infants would be necessary to support their postpartum growth. Higher fat reserves in infants would have been important to improve their chances of sur-vival during times when food was unavailable. So, the lyrics of Fred Rose's song "Roly Poly," about "daddy's little fatty," which was famously performed by Bob

---

[2] B. Pawłowski and A. Żelaźniewicz, "The Evolution of Perennially Enlarged Breasts in Women: A Critical Review and a Novel Hypothesis," *Biological Review* 96 (2021): 2794–809.

[3] C. W. Kuzawa, "Adipose Tissue in Human Infancy and Childhood: An Evolutionary Perspective," *American Journal of Physical Anthropology* (Suppl) 27 (1998): 177–209.

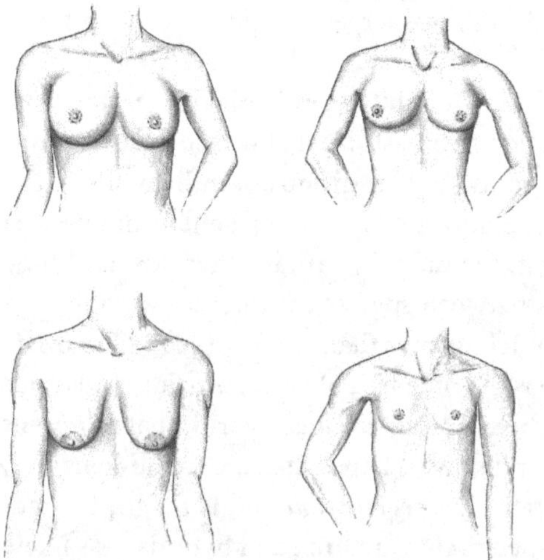

**Figure 6.3** Variations in human female breast size. Katherine Wonder (own work).

Wills and His Texas Playboys, are spot-on for describing our infants who bear the legacy of the challenges that our ancestors faced.

At this point, you're probably thinking of another question: if fat deposits were so important for infant survival in our *Homo* ancestors, then why don't all women have very large breasts (Figure 6.3)? We know that over the past 70,000 years our ancestors have colonized different habitats around the world that have varied in the reliability of food. In habitats where food availability was unreliable, we predict that there was selection to increase breast size and body fat deposits in general because infants of females with more fat would have had higher chances of survival. But selection would also work in the opposite direction—to reduce breast size—in regions where food was regularly available. Remember that all that fat is energetically expensive, so there will always be selection to reduce it if there is no reason to maintain it. It's a bit like a tug of war with selection pushing in both directions. Since fat deposits in women were closely tied to infant survival and reproductive success, selection would have been very strong, so responses over generations would have been relatively rapid. If people were living in a place where food availability was unreliable, then selection for increased fat reserves would have been stronger and breasts would have become larger over generations. But if people were living in a place where food was always available, then selection to decrease fat reserves would have been stronger and breasts would have become smaller. Previous authors have only focused on selection for larger breasts,[4] but understanding the range of selective forces affecting

---

[4] A. F. Dixson, *Sexual Selection and the Origins of Human Mating Systems* (Oxford University Press, 2009).

breast size helps us appreciate that this trait was largely an outcome of the types of habitats our ancestors lived in.

But variation due to selection in different habitats probably does not tell the whole story. In fact, the range of variation in breast size that we see today is probably greater than among our ancestors. Whenever groups of individuals of the same species are isolated for many generations, they develop genetic differences among groups. When they interbreed, the offspring are like hybrids, and they can be quite variable. When hybrids between species are formed, we often see a range of variation in traits that is much greater than what we see in the original genetic lineages. A good example is the breeding of irises, orchids, and other flowering plants. Plant breeders have been able to produce hybrids that represent a wider range of color combinations and flower shapes than are found in nature. This phenomenon is known as *transgressive segregation* and is due to interactions among genes from the parent species that are mixed in hybrids.[5] It's likely that genetic interactions have contributed to the wide variation we see today in breast size and shape, as peoples from different parts of the world have moved around and interbred.

We can be pretty confident that human breast size and other fat deposits in women represent adaptations to the long period of breastfeeding that our ancestors had to endure to ensure the survival of their offspring. Extended periods of breastfeeding became necessary as head size increased, but this was not an issue for our australopithecine ancestors. Remember that the heads of Lucy's species were one-third the size of modern humans', and calcium isotope analysis shows that they breastfed their infants for only a few months. We don't know for sure, but it's very unlikely that females of Lucy's species had enlarged breasts and other fat deposits, which probably developed in later hominins in the genus *Homo* as their infants required years instead of months of breastfeeding. We previously concluded that Lucy's species was likely to be nearly or completely hairless for cooling efficiency on the hot, dry savanna. It's notable that many artist illustrations depict Lucy as being much more apelike, with heavy coats of hair, and sometimes with breasts. But these illustrations are ill-informed and give false impressions about our australopithecine ancestors. It is more likely that Lucy was hairless, like modern humans, but she lacked fat deposits, so she would have been flat-chested, like female chimpanzees.

Now we can start making connections among more of the human traits we have discussed so far. As head size increased and our *Homo* ancestors' offspring were born earlier and earlier in development, there was strong selection to ensure the survival of infants. There is a coupling effect here; as mothers had

---

[5] M. B. Cruzan, *Evolutionary Biology—A Plant Perspective* (Oxford University Press, 2018).

to invest more time in breastfeeding to improve the chances of their infants' survival, selection on characteristics that increased the survival of infants became even stronger. For animals such as humans and elephants that have infants requiring parental care for many years, the survival of each one is that much more important. Compare that to animals like sea turtles, snakes, alligators, and insects that may produce dozens or even hundreds of offspring each year and provide little or no parental care. The result of selection for our ancestors to improve the chances of their offspring's survival was increased fat deposits in women and, as we discussed in the previous chapter, selection for cooperation, monogamy, and the inclusion of helpers such as homosexual individuals in clans.

We can conclude that the other ideas, such as the one that says women have large breasts just because heterosexual men prefer them or the one that proposes that women have large breasts to allow infants to breathe while suckling, are not supported by the evidence, or by any plausible argument. But maybe we can say that women have large breasts because humans have big heads? This sounds a little silly, but it's basically true. We have learned something important about how different habitats that our ancestors lived in may have selected for different breast sizes. We can guess that selection will always favor smaller breast size because large breasts are energetically expensive and represent a burden that could reduce survival. But when access to food is unpredictable, there will be strong selection to increase breast size and female fat deposits in general. Now you can appreciate that the shapes of our bodies represent outcomes of selection and genetic diversity in our ancestors; large breasts and buttocks, and the thicker fat layer under the skin of women, represent adaptations to improve the chances of offspring survival in our *Homo* ancestors and are not there just to please their partners.

*Summary*—In this chapter we discussed several hypotheses for the origin of bulbous breasts in females of our hominin ancestors. We learned about vestigial traits and concluded that large breasts in human females must have been favored by selection because otherwise breasts would have been reduced in size because they are energetically expensive. While we recognized that breasts are objects of sexual interest in many cultures around the world, this is not universally true, and mate choice is not a primary driver of selection for large breasts (but may have been a secondary driver, which is discussed in the next chapter). We came to this conclusion because we recognized that if there had been mate preferences for women with larger breasts, then breasts would be consistently large. This led us to explore the reasons for variation in breast size. First, we concluded that there would be selection for large breasts and increased fat deposits in the buttocks of our ancestors in environments where food supplies were unreliable. This is a consequence of the increased length of time that women had to breastfeed their infants. As head size increased in our *Homo* ancestors, infants became increasingly altricial and required several years of breastfeeding instead of just a few months, as was the case in our australopithecine ancestors. Consequently, infants of females with more fat deposits would have had higher chances of survival. By contrast, in environments where food supplies were more reliable, large breasts would be a liability because of the energy investment and decreased agility. Consequently, we would expect women with smaller breasts to have had higher chances of survival, and selection would result in decreased breast size. We noted that our babies have a high proportion of their birthweight as fat because of the energetic demands of their brains, which undergo rapid increases in volume after birth. We speculated that the range of breast sizes and shapes we see in modern societies is greater than in our ancestors as a consequence of transgressive segregation after interbreeding among people from different ancestral regions.

# 7

# Among the clans

*Launua's clan lived along with other clans in the same region, and it was not unusual for them to run into each other as they hunted or gathered food. While some clans had attacked and tried to steal food, weapons, and even women and children, more often these encounters were friendly and resulted in an exchange of skills and knowledge of the land. It was during one of these encounters that Launua's clan learned how to prepare food from the seeds of a grass that was common in the region. It required gathering large bundles of grass and then beating the straw to release the grain before it was ground into a powder, mixed with water, and cooked on hot stones. It took some effort to prepare, but the final result was quite tasty and had become a favorite among the clan members. On this day, Launua had left her children in the care of others as she worked in the savanna to gather grass into bundles. As she worked, the field of grass reminded her of earlier days in her young life when she first became aware of the young*

*Looking Down the Tree*. Mitchell B. Cruzan, Oxford University Press. © Oxford University Press (2025).
DOI: 10.1093/9780197805190.003.0007

*men in her clan as potential partners. These were the same boys she had grown up with and had known all her life, but now she was seeing them differently. There was one boy she felt most comfortable with. He was a good hunter. He was not the tallest or strongest, but he was well respected by his peers, and even the elders had taken notice of his prowess. They began spending more time together, and she had her first intimacy with him one day when they were walking out in the savanna by themselves. But there were other boys Launua's age, and one of them had begun to show some interest in her. Launua appreciated that he was the tallest and strongest boy in her clan, but he was not as good a hunter as her close friend. She did not feel as comfortable with this boy, but one day she gave in as he persisted in inviting her to walk in the savanna with him. She remembered looking back over her shoulder to see her close friend watching them as they left the camp. When she got back, he confronted her and seemed agitated—he wanted to walk with her again. During their intimacy, she was a bit alarmed at how urgent and forceful he seemed, but afterward they spent some time hugging and caressing. It was in that moment that she decided he would be her partner. As they lay in the tall grass together in the warm sun, she began to imagine their future life together. It was a beautiful memory and one of the reasons she enjoyed working in the grasslands—it reminded her of some of the best moments of her life.*

We have all seen videos and pictures of the spectacular mating displays of birds and other animals. Some examples you may have heard about are grouse and prairie chickens in western North America. Males of these species inflate large air sacs on their throats while making guttural sounds as they jockey for position at the center of the display arena called a *lek*. Females enter the lek and usually choose to mate with the males that have successfully fought off others to maintain their position in the center. Males of the magnificent frigatebird inflate bright red throat pouches and drum on them by clacking their bills if they see a female fly overhead. Peacocks are another example you are probably familiar with; males have spectacular tails that they display, while pea hens are brown with small tails. The same is true in many bird species, where the males are brightly colored while the females are much drabber. You may also know that male elk, deer, and other large mammals have antlers or horns, while the females don't. They display themselves to females by engaging in ritual battles with other males while the females judge them and decide which male to mate with. In all of these cases, it is the male that has the brighter color or elaborate displays and mating behaviors, while the females are drab and less ornamented. Males display and females choose based on their appearance and performance.

This was one of the puzzles that Charles Darwin had to solve; why are males of many animals more colorful and ornamented compared to females? Darwin proposed that these displays only improve mating success and are not favored by natural selection because they do not improve survival; in fact, they often reduce the survival of males. He used the term "selection in relation to sex," which we now call *sexual selection*, to describe the process of evolution for improved mating success as the primary reason for the evolution of these displays. Did sexual selection affect our hominin ancestors? We want to understand whether sexual selection has affected human evolution, and to do that we need to more closely look at examples from other animals so we can better understand how sexual selection works.

The first question we need to ask is, how do spectacular male displays evolve? One idea is called *runaway selection*, where females prefer males with a particular kind of display—say a longer tail—and then their sons have the same characteristics, which in turn improves their mating success and, consequently, the fitness of their mothers. This idea, which is called the *sexy son hypothesis*, was assumed to be the primary reason for spectacular male displays in many species for many years. But in the 1980s, mathematical models showed that this process of runaway selection would not work because it didn't include a mechanism to change female preferences for more and more exaggerated displays.[1] In fact, the models predicted that tail length was just as likely to decrease as it was to increase. Without something driving female preference for longer and longer tails, the idea of runaway selection comes to a halt. What is causing the evolution of these spectacular mating displays in males?

The answer to this puzzle is that innate biases of females probably drive sexual selection; females are "hardwired" to prefer males that have exaggerated mating displays. This type of innate bias seems to be common. A good example is fish in the genus *Xiphophorus*, which includes species of platies and swordtails, which you may be familiar with because they are commonly kept in aquariums. Males of the swordtail fish have elongated tails, while females do not. The elongated tails are the result of sexual selection due to female preferences for males with longer tails. Experiments have shown that in females the innate bias for long tails is even present in the sister species, the platy fish, which do not have sword tails.[2] The experimenters discovered this by gluing plastic swords onto male platy fish, and sure enough, the females preferred the males with plastic swords.

---

[1] R. Lande, "Models of Speciation by Sexual Selection on Polygenic Traits," *Proceedings of the National Academy of Sciences* 78 (1981): 3721–25.

[2] A. L. Basolo, "Female Preference Predates the Evolution of the Sword in Swordtail Fishes," *Science* 250 (1990): 808–10.

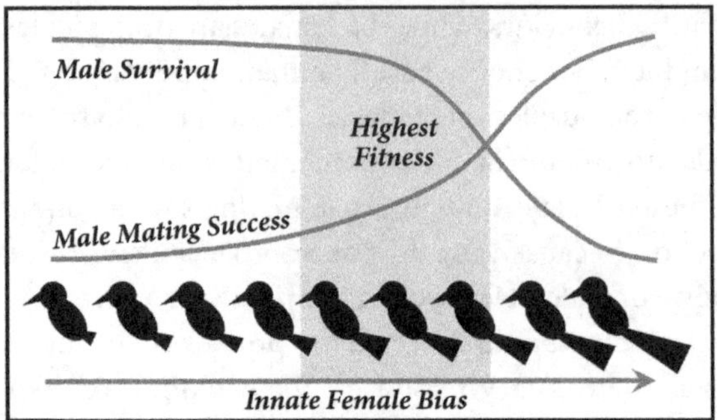


**Figure 7.1** The effects of innate female bias on runaway sexual selection for male tail length. Male mating success increases and survival decreases with increasing tail length. M.B. Cruzan (own work).

In another study with the túngara frog from Panama, researchers were able to determine that male calls that were longer and more complex increased stimulation and excitation in the brains of females.[3] Longer calls also decreased fitness because these males were more likely to be subjected to predation by bats. So sexual selection can drive the exaggeration of traits in males, but it can be opposed by natural selection if the trait lowers the chances of survival, as shown in Figure 7.1. These results support the idea that sexual selection for exaggerated male traits in many species is probably driven by innate female preferences, which can even be shared among closely related species.

So why are males susceptible to sexual selection and females are not? Think about the time and energy that males and females of different kinds of animals contribute to reproduction. In the extreme case, males only contribute sperm for fertilization, while females are responsible for producing eggs or live offspring and often for their care and feeding—the difference in the resource investment is dramatic. This idea is referred to as *Bateman's principle* and was proposed by Angus Bateman in the early twentieth century. Since male fitness is only limited by how many females they can mate with, there should be strong selection to improve their chances for mating with as many females as possible. The variation in male mating success is high, and males with more exaggerated displays often have higher mating success, which is why we see these amazing colors and displays in males of many species. Meanwhile, females are saddled with making sure offspring have enough food and care to become independent, so their

[3] M. J. Ryan and A. S. Rand, "The Sensory Basis of Sexual Selection for Complex Calls in the Túngara Frog, *Physalaemus pustulosus*," *Evolution* 44 (1990): 305–14.

fitness is limited by access to resources—meaning food. Bateman's principle predicts that sexually selected traits, like elaborate colors and displays, should only be found in the mate-limited sex, which is usually males.

Seahorses are the "exception that proves the rule" when it comes to sexual selection. Male seahorses produce sperm, but the female transfers her eggs to a pouch on his belly. Males fertilize the eggs and are then responsible for caring for the developing young until they can feed themselves, so they have to make a much larger investment in their offspring than males of most species. Knowing about Bateman's principle, what would you predict about which sex should be choosier about who they mate with? In seahorses, the gender roles are switched. Since the males are the ones making the big investment in terms of the time required for care of offspring, they are choosier about who they mate with, and they prefer to mate with females that produce more eggs.

To understand which human traits might have been subject to sexual selection in our ancestors, let's look at the history of ideas on this topic, starting with Charles Darwin. In his book *The Descent of Man and Selection in Relation to Sex*, Darwin identified several human traits that might be best explained by sexual selection. These include hairlessness in women and beards in men. Darwin erroneously thought that hairlessness in humans was due to males preferring women with less hair and that the loss of hair in men was a correlated response to sexual selection on women. We already noted that loss of fur in our ancestors is best explained by the need to keep cool by sweating, but we did not discuss the difference in body hair between men and women.

Darwin's view on body hair differences between men and women was clouded by his limited experience with peoples from around the world. What he and some more recent authors did not appreciate is that people from East Asia and Africa are mostly hairless, so there is little difference in body hair between men and women. The difference is greater in people from Europe, where men tend to have more body hair than women. It's a similar story with beards—men from Asia often only grow hair on their upper lip and chin, and not on their cheeks. These patterns are a bit muddled nowadays because almost everybody has some mixed ancestry, but the general pattern appears to be that men from western Asia and Europe have full beards and more body hair than women, but people from other parts of the world are a lot less hairy. Nonetheless, among people of European ancestry, beards are often associated with social status[4] so the evolution of facial hair in men may have been affected by mate-choice preferences in those cultures.

---

[4] C. C. Grueter et al., "Are Badges of Status Adaptive in Large Complex Primate Groups?," *Evolution and Human Behavior* 36 (2015): 398–406.

The problem we have with assuming that male body hair and beards are due to sexual selection is that they could have been favored because our Eurasian ancestors—and Neanderthals in particular—lived in a very cold environment. In that case, it could be that beards and body hair were adaptations for keeping warm during harsh winters. Perhaps there was a female preference for beards and body hair in our ancestors in northern regions of Eurasia, but it's not likely that this female preference was universal. It's interesting to note that the hairiest people in the world are the Ainu, who appear to have originated from the northern regions of Eurasia,[5] so it seems that there is a good association between hairiness in people and originating from northern latitudes.

It's clear that sexual selection has been important in other species, so what traits might have been affected in our ancestors? We know that any trait that is different between males and females might be under sexual selection. But keep in mind that we are discussing traits that would have been different in our ancestors, so we can't include things that have to do with grooming, like hair styles and hair removal. We already discussed facial and body hair in men in this chapter, as well as breasts and buttocks size in women in the previous chapter, so what other characteristics may have affected by sexual selection in our hominin ancestors?

Before we proceed with our discussion of sexual selection in humans, we need to decide which is the mate-limited sex; did males or females of our *Homo* ancestors invest more in the raising of offspring? Of course, we would say females invested much more because they are the ones who carried the developing infant and nursed them. But does that mean that male or female partners that were not breastfeeding didn't invest much at all? Remember we discussed the importance of parental care for offspring survival as brain size increased in our *Homo* ancestors, and we predicted that there would have been increasing selection for paternal care as infants became increasingly altricial. It's interesting to note that some authors insist on portraying human males as promiscuous and mating with as many females as possible to increase their fitness.[6] This view comes from male behavior in contemporary societies and does not account for the fact that paternal involvement in childcare was crucial to ensure the survival of their offspring in our *Homo* ancestors. Mating with many females would not have increased their fitness if the resulting infants had low chances of surviving.

Extended periods of parental care were important for the survival of offspring in our ancestors, especially when they were living in harsh environments. Because of the increased investment in childcare, we expect that males as well

[5] T. Sato et al., "Whole-Genome Sequencing of a 900-Year-Old Human Skeleton Supports Two Past Migration Events from the Russian Far East to Northern Japan," *Genome Biology and Evolution* 13 (2021): evab192.

[6] J. Diamond, *Why Is Sex Fun?* (Basic Books, 1997).

as females of our *Homo* ancestors might have been choosy about who they mated with. Once a male fathered an infant, it was a long-term investment to make sure the child would grow into an adult. If they had mated and then abandoned their family, the chances of the children's survival would have been a lot lower, so that type of behavior would not have been favored by selection. Once males needed to invest more time in childcare to increase their fitness, they might have become choosier. We can predict that in environments where food availability was unreliable, males would have preferred females with larger breasts and buttocks because their children would have had a higher chance of survival. This was not a conscious choice in our *Homo* ancestors but would have been an innate bias that had been favored by natural selection. Just like the seahorse example we discussed above, based on Bateman's principle, selection would have favored males who were choosier about who they mated with because of their increased investment in offspring. Natural selection was the primary driver favoring larger breasts and buttocks in females of our *Homo* ancestors who lived in habitats where food sources were less reliable, but it is likely that mate choice by males—sexual selection—also favored these characteristics.

What about other human traits? A number of evolutionary biologists have speculated that the relatively large size of the human penis compared to other great apes is a consequence of sexual selection by female choice, while others have argued that human penis size is not unusually large compared to other primates.[7] The idea that sexual selection has favored larger penises assumes that there is universal and consistent female preference for large penises. However, surveys suggest that penis size is not a high priority for most women and that preferences favor penises that are only slightly larger than the average.[8] In fact, women are much more likely to choose mates based on social status.[9] What about the evolution of penis size in other mammals—is there evidence for sexual selection? In some rodent species there is, though not in primates.[10] But we still have not answered the question: what drove the evolution of large penis size in humans?

One pattern we can find across mammals is a strong correspondence between vagina length and penis length. The argument for this relationship is that there is strong selection to ensure that sperm is deposited as close to the cervix as possible to increase the chances of fertilization. Maybe we're thinking about it

---

[7] A. F. Dixson, *Sexual Selection and the Origins of Human Mating Systems* (Oxford University Press, 2009).

[8] N. Prause et al., "Women's Preferences for Penis Size: A New Research Method Using Selection among 3D Models," *PLoS One* 10 (2015): e0133079.

[9] N. P. Li et al., "Mate Preferences Do Predict Attraction and Choices in the Early Stages of Mate Selection," *Journal of Personality and Social Psychology* 105 (2013): 757–76.

[10] S. A. Ramm, "Sexual Selection and Genital Evolution in Mammals: A Phylogenetic Analysis of Baculum Length," *American Naturalist* 169 (2007): 360–69.

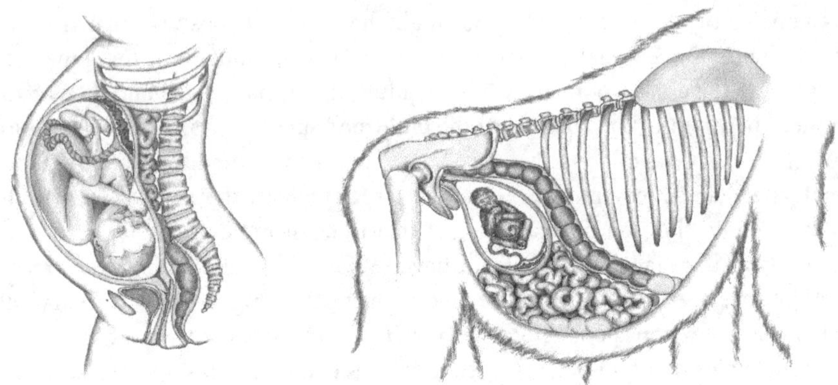

**Figure 7.2**  Anatomy of a pregnant human and gorilla. Katherine Wonder (own work).

the wrong way; if selection on penis size is just based on vagina length, then perhaps we should be looking for what caused longer vaginas in humans.

To understand why human vaginas are longer than other primates, we need to remember that the vagina is not just for sex—it also functions as the birth canal. If we look at anatomical drawings of pregnant primates, we can find some clues to solve the puzzle of why humans have long vaginas (Figure 7.2). It's interesting to note that it is difficult for zoo veterinarians to determine when gorillas are pregnant because they don't "show" nearly as much as pregnant humans. If you look at Figure 7.2 (right), which shows a pregnant gorilla, you will see the tiny baby nestled down between the hip bones. The baby gorilla, with its relatively small head, can fit nicely in the lower abdomen, and consequently the gorilla vagina is only about two inches long. As you might expect, an erect gorilla penis is only about two inches long as well (Figure 7.3, middle), whereas human penises average about five to six inches long when erect (Figure 7.3, right). Now, contrast that with the anatomical illustration of a pregnant human (Figure 7.2, left); the head of the baby is so large that it would not fit easily in the lower abdomen like in the pregnant gorilla. Instead, the baby is positioned above the hips and much higher up in the abdomen than in gorillas. Consequently, the vagina is proportionally longer to accommodate the higher position of the uterus. This pattern is consistent with what we expected from comparisons of penis and vagina length across mammals; selection favored a longer vagina as head size increased from our *Australopithecus* ancestor to *Homo erectus* and later *Homo sapiens*, and consequently there was selection for a correspondingly longer penis.

You mean humans have both big breasts and big penises because we have big heads? Again, that sounds silly, but yes, all of the evidence we have discussed lines up to support that conclusion. But the mechanism of selection

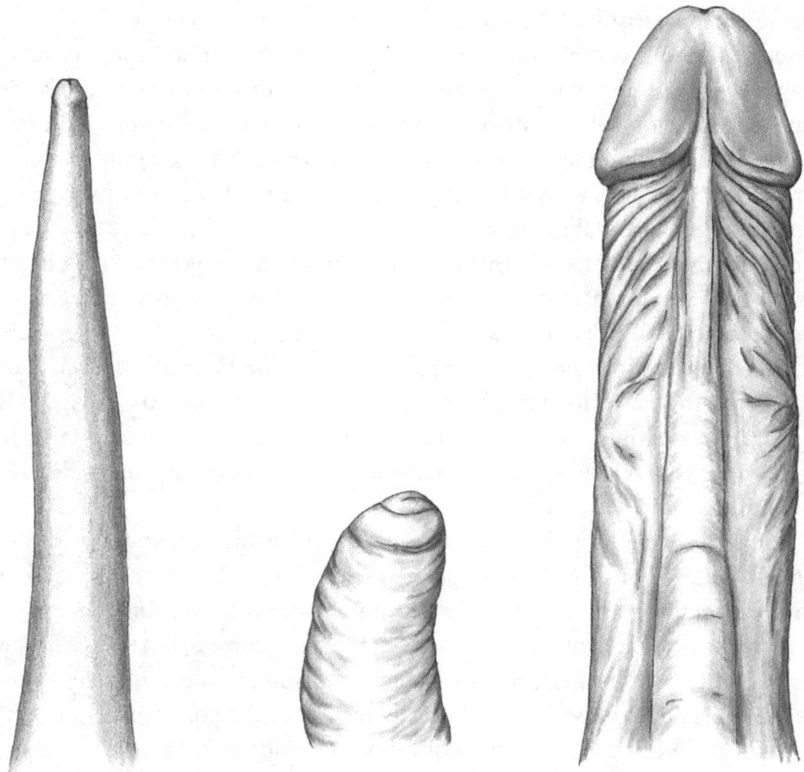

**Figure 7.3**  Morphology of chimpanzee, gorilla, and human erect penises. Katherine Wonder (own work).

differs between these traits. Males may have preferred to mate with women with larger breasts, and even though there may have been some role for sexual selection for breast size, the primary reason for larger breasts—and generally for fat deposits in human females—is selection to improve survival of offspring in environments where food availability was unreliable. As for the penis, there isn't any evidence that female preference would cause sexual selection for increased size. There is strong evidence across mammals for selection to deposit sperm close to the cervix; consequently, we can conclude that longer vaginas caused selection for increased penis length.

If that's true, then why do chimps and bonobos have such long penises? You can see in Figure 7.3 (left) that an erect chimpanzee penis is about five inches long and much thinner and narrower than a human penis. They are shaped like a carrot and come to a point at the tip. They also lack the foreskin and glans that are characteristic of the human penis. Given the smaller heads of chimps, we can assume that their internal anatomy should be similar to the gorilla. So why is it so long? If you look up more information on chimpanzee and bonobo reproduction, you will likely run across images of female chimpanzees in estrus. Unlike humans, chimpanzees do not ovulate on a monthly cycle but may become fertile (i.e., they are in *estrus*) at any time of the year. When in estrus, females have thick genital swellings that are pink in color to signal to males that they are fertile. It's interesting to note that the primate with the longest penis for its body size is the Hamadryas baboon. They are smaller than chimpanzees, and their erect penis is about five inches long. Like the chimpanzee, their females have large genital swellings when they are fertile. These swellings effectively make the distance to the cervix longer, and consequently, the penis has to be longer. Once again, it's one of those exceptions that proves the rule; selection on penis length to deposit sperm close to the cervix works in humans, chimpanzees, bonobos, and other primates as well.

But now we have another puzzle to solve; why is the human penis so thick? The diameter of an erect human penis is several times the diameter of a chimpanzee's penis. You should note that the human penis has a *glans* and a *foreskin* that are not present in our closest relatives, the chimpanzees, but the gorilla penis has a vaguely similar morphology.[11] While our hominin ancestors developed a penile morphology that is an elaborated version of the gorilla penis, the chimpanzees' penis became much more filiform (long and thin). As we have discussed before, such structures do not come about by accident, so there must be some evolutionary motivation that led to the origin of a thick penis with a glans and foreskin since we diverged from gorillas and chimpanzees.

To understand the origin of the characteristics of the human penis, it is useful to look for other species that have similar elaborate characteristics of their copulatory organs. For example, many insect species have elaborate structures on the aedeagus, which is the insect equivalent of a penis. In those cases, the elaborate morphologies are mechanisms to remove the sperm deposits of a previous male, giving the male that mated most recently the advantage of fertilizing most of the eggs. This is a phenomenon referred to as *sperm competition*, where the forum for selection takes place within the reproductive organs of females. It is an example of *sexual conflict*—when the goals of the male and the female are not the same; the male intends to fertilize all of the female's eggs, whereas the female would have higher fitness if her offspring had higher genetic diversity by bearing offspring fathered by multiple males.

[11] A. F. Dixson, *Sexual Selection and the Origins of Human Mating Systems* (Oxford University Press, 2009).

The sexual conflict hypothesis assumes that females will benefit by increasing the genetic diversity of her offspring. High genetic diversity increases the chance that some of a female's offspring will have high fitness—sort of like a lottery where you need the right ticket to win. If she has children with multiple males, then there's a better chance that some of them will have the right combination of genes to increase their survival and reproductive success. An individual male, on the other hand, wants to father all of her offspring—that is the best way to increase his fitness. Male strategies to ensure that they father as many offspring as possible include prolonged copulation, which excludes other suiters and occurs in some insects; the vaginal plug deposited by male mice to prevent copulations with other males; and structures on the male reproductive organs of insects that remove sperm deposits from previous males. These mechanisms reduce the chances that a particular male's sperm will be in competition with sperm from other males to fertilize eggs—they reduce sperm competition.

The human penis is unique compared to other primates because it is much thicker and has a glans and foreskin. Studies have shown that the combination of these characteristics is effective in removing the sperm deposits of a previous male via thrusting during copulation.[12] It's true that humans have much more energetic pelvic thrusting than most mammals, and studies have shown that men report more intense thrusting if they know or suspect that their partner has recently been with another man. Some authors have dismissed this idea[13] but have not offered alternative explanations for human penile morphology. It's possible that the morphology of the human penis is simply a holdover from an ancestor; the human penis is perhaps just an enlarged version of a gorilla's penis, and apparently the chimpanzee penis has deviated in shape compared to our common ancestor.

On the other hand, genomic analyses have found that gorillas, which live in polygynous units and therefore are not subject to sperm competition, have lost a number of genes responsible for sperm development and function.[14] These genes are associated with sperm competition and have been maintained in both human and chimpanzee genomes. Chimpanzees are known to be subject to sperm competition because they live in multimale/multifemale groups, and females in estrus engage in copulation with several different males. The observation that our ancestors retained genes important for sperm function

---

[12] G. G. Gallup and R. L. Burch, "Semen Displacement as a Sperm Competition Strategy in Humans," *Evolutionary Psychology* 17 (2006): 253–64.

[13] A. F. Dixson, *Sexual Selection and the Origins of Human Mating Systems* (Oxford University Press, 2009).

[14] A. Scally et al., "Insights into Hominid Evolution from the Gorilla Genome Sequence," *Nature* 483 (2012): 169–75.

suggests that sperm competition has been important in human evolutionary history too, which is consistent with the idea that the human penis morphology evolved to displace sperm deposits of previous males from the vagina. Because our ancestors lived in multimale/multifemale groups, females may have mated with multiple males over a short period of time. In chimpanzees, the effects of sperm competition led to selection for the production of large sperm deposits, and consequently, they have the largest testicles relative to body size among primates.[15] Remember that complex structures like the glans and foreskin are not maintained by accident—these structures must have been favored by selection. A logical explanation is that they have been favored to remove sperm deposits by other males from the vagina. This tells us that females of our ancestors often mated with multiple males over short periods of time.

I know that this idea is very foreign to us, that the mating behavior of our ancestors was so different from acceptable norms of modern societies. But it's important to remember that our ancestors were very different from modern humans. The available evidence tells us that a human penis morphology that reduces sperm competition may have been favored in our ancestors living in communal situations where multiple males mated with the same female.

How did we come to this conclusion? Let's review the evidence. First, remember that the human penis is much thicker and has a glans that may have been carried over from a common ancestor with gorillas. These structures have been elaborated by selection in our hominin ancestors who lived in groups and engaged in mating with multiple individuals over short periods of time. Second, selection has maintained genes in humans that are associated with sperm competition. These same genes have been maintained in chimpanzees, which are known to be subject to sperm competition, but they have been lost in gorillas, which are not subject to sperm competition. Third, we know about analogous structures in other animals and their role in removing sperm deposits from previous males and consequently reducing sperm competition. And fourth, we have experimental evidence that these structures are effective in removing semen deposits from the vagina. It's hard for us to imagine today, but multiple copulations and sperm competition must have been common; otherwise, we would not be seeing the maintenance of genes associated with sperm competition and this complex penis morphology in modern humans.

But that does not answer all of our questions about human penis morphology. We also need to discuss another difference between humans and other primates: the loss of the penile bone—the baculum. Humans are one of the few primates, and the only great ape, that do not have a penile bone. Whenever we see a trait

<hr>

[15] A. H. Harcourt and A. L. Purvis, "Sperm Competition: Mating System, Not Breeding Season, Affects Testes Size of Primates," *Functional Ecology* 9 (1995): 468–76.

in the majority of members of a phylogenetic group, it means that the trait was most likely present in their common ancestor. Based on the phylogeny of the great apes, we can conclude that the baculum was present in our common ancestors with chimpanzees and gorillas. One explanation for the loss of the penile bone is sexual selection. Some authors have speculated that without a baculum, it is more difficult for an unhealthy male to maintain an erection, so this would be a mechanism favored by sexual selection for females to assess male vigor and health.[16]

Does it make sense that loss of the penile bone is due to sexual selection? It seems to make sense that females choose healthy males, right? But remember that the peacock's tail and other traits subject to sexual selection like antlers and bright colors in birds are all getting bigger or brighter. We see many examples of exaggerated traits in males—longer tails, larger displays, and brighter colors—but there aren't any examples where something was lost due to sexual selection. This hypothesis does not explain why the penile bone wasn't lost in the other great apes. Consequently, we can conclude that the "male health" hypothesis for loss of the baculum do to sexual selection does not make sense in the context of mate choice in our ancestors.

The answer to the loss of the baculum has more to do with penis size, especially erect penis size. In some respects, humans are more like horses and other ungulates, which also lack penile bones. Note that these animals have very large penises when they are erect so they can reach the cervix when mounting a female. But their penises are very small when flaccid, which is probably necessary to improve their ability to run to escape predators, and in the case of humans, to also be a predator.

Think about how it would be if these animals did have a penile bone—it would have to be very large to be functional at all, and that would put a constraint on how small the penis could be when flaccid. This is another case where we can learn something about human evolution by comparing ourselves with animals that have similar traits. For both horses and humans there was selection to increase penis length, and at the same time, there was selection for walking and running over long distances.

It makes sense that the loss of the penile bone in humans had to do with our transition to the savanna and the increase in brain size in our ancestors. As there was selection to increase penis length as vaginas became longer in our *Homo* ancestors, there was also selection for the ability to run long distances and move more efficiently for hunting and escaping from predators. The solution was the same as what must have happened in the common ancestor of the

---

[16] R. Dawkins, *The Selfish Gene* (Oxford University Press, 2006).

ungulates: to lose the penile bone and instead rely on hydrostatic pressure—blood engorgement—to maintain an erection. We can conclude that the penile bone was lost to improve running efficiency as there was selection for a longer penis, and we can be pretty sure that it was not just to allow females to assess male health.

The diversity of shapes among primate penises shows us that penis morphology is often subject to selection. The simplest and best explanation for this diversity of morphologies across primates—including humans—is that there has been strong selection to improve the chances for successful fertilization. Selection has favored a penis length to deposit sperm as close to the cervix as possible, and it has favored the development of characteristics in the human penis that would reduce sperm competition from other males.

*Summary*—In this chapter we learned how innate female preferences in many species have led to elaborate male behaviors and displays and bright coloration through sexual selection. Bateman's principle predicts that individuals will be choosy about who they mate with when they have to invest more resources to ensure the survival of their offspring. This led us to the conclusion that selection for increased paternal childcare in our *Homo* ancestors would have resulted in selection for male mate preferences favoring females having more fat deposits, but only in environments where food availability was unreliable. Consequently, natural and sexual selection worked in concert to affect the evolution of fat deposits in females of our *Homo* ancestors, resulting in larger breasts and buttocks in challenging environments and smaller breasts and buttocks when food availability was more reliable.

We examined other differences between males and females including facial and body hair. While Darwin and others thought that the loss of hair in females and beards in men was an outcome of mate choice preferences in our ancestors, we concluded that the evidence for this is weak. It's just as likely that selection favored more hair for populations of our *Homo* ancestors living in colder regions of Eurasia.

There has been speculation that penis size in humans is an outcome of female mate choice, but there isn't any evidence available to support this hypothesis. Instead, we realized that the increase in head size in our *Homo* ancestors required that the uterus move higher into the body cavity, resulting in a longer vagina. Since there's strong selection on penis length to deposit sperm close to the cervix, penises became longer as head size increased in our *Homo* ancestors.

We learned about the concepts of sexual conflict and sperm competition. Sexual conflict can occur in species that birth multiple live offspring or eggs because there is selection for females to ensure the genetic diversity of their offspring by mating with multiple males, while there is selection for each male to fertilize as many of a female's eggs as possible. Sperm competition results in selection on males to improve their chances of fertilizing eggs. In chimpanzees, multiple males mate with fertile females, and sperm competition has resulted in selection for large sperm deposits; consequently, they have very large testicles. Genomic evidence indicates that sperm competition was important in our human ancestors and was probably responsible for humans' uniquely thick penis along with its glans and foreskin. This idea is supported by the retention of genes associated with sperm competition in humans and chimpanzees but not in gorillas. Penis morphology has been shown to be effective in removing a previous male's semen deposits from the vagina, which would have been important in our ancestors when multiple males mated with the same female.

Lastly, we rejected the hypothesis that loss of the baculum in our human ancestors was due to mate choice by females so that they would have an honest assessment of male health. Instead, we concluded that loss of the baculum was due to selection for efficient walking and running on the savanna combined with increased penis length, because an effective baculum would have to be very large and would have interfered with the ability to run fast and over long distances.

# 8

# The good life

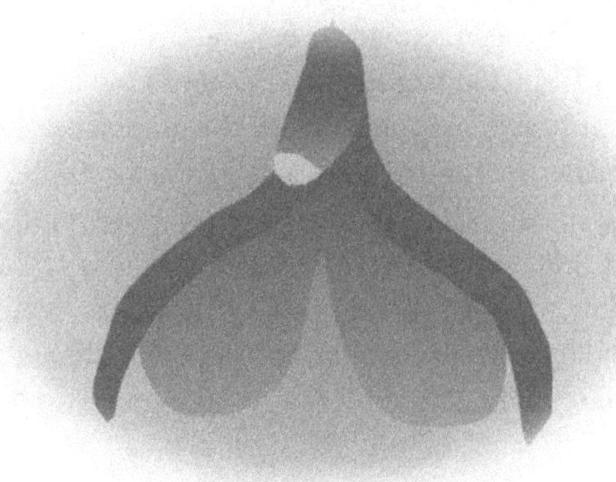

*Launua and her partner were very happy in their new home, and all of the clan members seemed to be in good spirits. Food was plentiful in this valley; the gatherers returned early each day with plenty of roots, and the men and women who went hunting were bringing in more meat than the clan could eat. Life was easy, so they had much more time to socialize and exchange stories during the long afternoons and over the evening campfires. They used new skills they had learned from other clans to clean and dry the skins from several of the large animals they had killed and stitch them together so every family had a tent to protect them from the cold night air. There was more time for leisure and playing with her children, and Launua was enjoying her life here. But her favorite times were late in the evening when the children were sleeping. She would listen to the sounds in the camp: occasionally a baby would cry, and she could hear the noises of other couples making love. It was at these times, when she felt warm and secure, that she would reach out to touch her partner, and he would respond by rolling over to hold her close. After their lovemaking, they would lie together.He had learned how to please her and she enjoyed his gentle caressing. He used his fingers to rub between her legs until*

*Looking Down the Tree*. Mitchell B. Cruzan, Oxford University Press. © Oxford University Press (2025).
DOI: 10.1093/9780197805190.003.0008

*her thighs tightened and she trembled as a rush of pleasure surged through her body. After her breathing quieted, they lay together and she would fall into a deep, comfortable sleep, knowing that her family was safe.*

We have already discussed how living species share many characteristics that have been inherited from common ancestors, like the fact that all mammals have mammary glands and hair, but what about traits that have been lost? You might remember our discussion in the previous chapter on vestigial traits. These are structures that have been lost or reduced because they are no longer useful—they do not improve the fitness of individuals. As we discussed before, one of the most dramatic vestigial traits is the remnant of hip bones in marine mammals and snakes, which have lost their hind limbs. But there are many vestigial traits in humans as well; for example, the tailbone, wisdom teeth, arrector pili muscles, and some small muscles within our cartilaginous ears. Note that the tailbone has been reduced in the great apes, but only to a certain point because it is still functional in its reduced state as it has muscle and tendon attachments. The arrector pili muscles were important in furry animals to allow them to "puff up" and appear larger when threatened. These small muscles are responsible for goosebumps in humans, but they don't have the same effect on our mostly hairless bodies as they did in our furrier ancestors. These vestigial traits are evidence of our ancestry and the common ancestors that we share with other species.

Once our ancestors started using fire to cook food, there was no longer selection to maintain a very large and muscular jaw since the food was softer and more easily digested. Cooking vegetables breaks down the cell walls and makes it easier to absorb the nutritional content, and cooking meat increases the amount of energy our digestive systems can extract.[1] Once our ancestors started routinely eating cooked food, mutations that reduced jaw size and heavy musculature were no longer opposed by selection, so there was a gradual reduction in jaw size. But not all hominins in Africa took advantage of fire. For example, the megadont hominins in the genus *Paranthropus* had very large jaws and teeth along with a cranial ridge across the tops of their skulls for the attachment of large muscles for chewing raw foods. If you are a Star Trek fan, you will be familiar with Klingons, a fictional humanoid species from the planet Kronos who routinely ate raw meat. The creators of Star Trek accurately portrayed Klingons with large cranial ridges, but they got the jaw wrong—a true raw-meat-eating species would have had much larger jaws and teeth than those portrayed on the Klingons in the earlier Star Trek series and movies.

[1] Carmody, R.N., et al. 2011. Energetic consequences of thermal and nonthermal food processing. PNAS 108: 19199–19203.

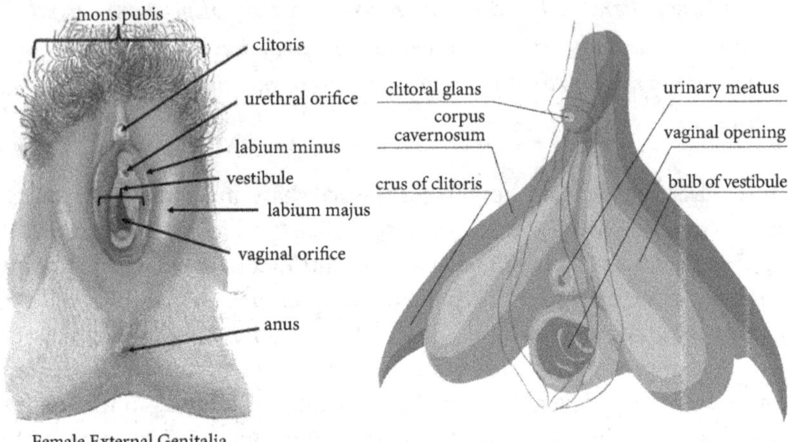


**Figure 8.1** External anatomy of the human vulva and internal anatomy of the clitoris.

Reproduced from Amphis (2024) via Wikimedia Commons and Cypressvine (2018) via Wikimedia Commons, Creative Commons Attribution-Share Alike 4.0 International (CC-BY-SA-4.0).

Once our ancestors started using fire to cook their food, large jaws and jaw muscles were no longer needed, but apparently the number of teeth in our jaws is evolutionarily constrained and not as susceptible to reduction, because most of us have the same number of teeth—thirty-two—as the other great apes. There are exceptions, like some natives of Central America who no longer have wisdom teeth.[2] For the rest of us, we often have our wisdom teeth surgically removed to make room for all the other teeth that are crowded into our small jaws.

One of the more interesting examples of human vestigial traits is the little muscles in our ears. These tell us that, at some point in the past, our ancestors were able to move their ears to change direction—the same way that cats move their ears. But our ears are rigid now, so these muscles are no longer needed; they have been reduced in size but not completely lost.

There are some features of the human body that superficially appear to be vestigial but on closer inspection are clearly not. As you can see in Figure 8.1, a women's genital area has labia, a urethra, a clitoral hood, and a clitoris, and these structures correspond to male genitalia. The hood is derived from the same tissue as the foreskin of the penis, and the clitoris is derived from the same tissue as the glans, so male and female genitalia have homologous structures; they share

---

[2] Main, D. 2013. *Ancient Mutation Explains Missing Wisdom Teeth*. Live Science (https://www. livescience.com/27529-missing-wisdom-teeth.html).

the same developmental origin. The question we want to explore is whether the clitoris is an artifact of development—does it have any real function, or is it like nonfunctional breasts and nipples in men? The one thing we know for sure is that it enables women to have orgasms, so maybe the real question we are asking is whether female orgasm had a function that improved fitness in our ancestors. For males, orgasm is necessary for ejaculation, so it has a reproductive function. But what about for women?

The origin and maintenance of the clitoris and female orgasm has been the subject of much discussion and debate since the early 1900s. There are a lot of sources of information both online and in books on human sexuality and evolution asserting that the clitoris is just a vestigial penis and has no adaptive function at all. The argument for this is that female orgasm is not necessary for successful fertilization, and it does not seem to serve any purpose. In contrast, breasts in men are fully formed with all the mammary glands they need for milk production—the only thing that is missing is the fat deposits that characterize adult female breasts.

Why is the clitoris so small and seemingly vestigial? On the surface, it seems like the clitoris is a miniature version of the penis and female orgasm doesn't seem to have any real purpose. One famous evolutionary biologist, Stephen Jay Gould, was of the opinion that the clitoris is a vestigial trait, and many others have repeated this idea.[3] But opinions are not meaningful in science unless they can be supported by data and logical arguments. It doesn't matter whether it is voiced by a famous biologist or even Charles Darwin; no opinion is worth anything unless it can be backed up with data. So, let's have a look at the available data.

In bonobos we know that female orgasm and sexual interactions in general are extremely important for social bonding within groups and for greetings between groups. Individuals engage in sex prior to eating whenever they find a good source of food, and there are lots of sexual interactions when two groups encounter each other. Many of these interactions are between females. Female bonobos often engage in mutual clitoral stimulation when embracing face to face. They can do this easily because their clitoris is larger and is shifted upward and forward from where it is located in human females. We can conclude that the clitoris and female orgasm are extremely important for bonobo social interactions, but that does not necessarily tell us anything about its function in humans. It's important to note that while female gorillas and chimpanzees are capable of orgasm, it is not important in their social interactions and may not occur very often during mating.

---

[3] Dixson, A.F. 2009. *Sexual Selection and the Origins of Human Mating Systems.* Oxford University Press.

But is the human clitoris just a nonfunctional, miniature penis? Absolutely not! The clitoris is a complex structure; it has been called an "electrochemical wonder" with more than 8,000 nerve endings packed into a small volume. Moreover, it can expand to three times its normal size when a woman is aroused. Even if you shrank a penis down to the size of a clitoris, the clitoris would still have a higher density of nerve endings. Remember, just as we said for the foreskin and glans of the penis, such complex structures do not come about by accident. If the clitoris had not been important for increasing fitness in our ancestors, then we can expect that it would have been lost over time and become truly vestigial. The only reasonable conclusion we can make is that selection has favored not only the maintenance but also the elaboration of the clitoris compared to the penis.

One idea that has been promoted by a number of people to explain female orgasm is that it improves the chances of fertilization and pregnancy. It's true that during orgasm there are convulsions in the vagina and uterus causing the cervix to dip downward, and some have proposed that this would aid in the uptake of sperm into the uterus. Clitoral stimulation causes physiological changes in the vagina to activate sperm and facilitate fertilization,[4] but these changes happen without the occurrence of an orgasm. It has been suggested that after an orgasm, women tend to lie still, which improves sperm retention in the vagina. This seems like a good idea, but unfortunately, there isn't evidence available to support the hypothesis that a woman's chances of pregnancy are improved if she has an orgasm during or shortly after copulation.[5] Female orgasm is not necessary for successful fertilization, so it is unlikely that improving the chances of fertilization is the primary function of female orgasm.

Others have suggested that maybe our ancestors were more like bonobos. But remember that the human clitoris is much smaller than in bonobos, and it's positioned wrong for easy mutual genital rubbing. Also, our social interactions don't depend on sex like they do in bonobos. If we were like bonobos, then every time you met up with a friend you hadn't seen in a while you would have sex on the spot, and there would be a lot of sex going on in public. But maybe our ancestors were more like bonobos and the clitoris moved downward when they became bipedal? That's possible, but selection would have continued to favor a frontal position if that had been important for social interactions, and that does not explain why it became smaller. Perhaps the clitoris is important for orgasm during copulation? But it's position is not optimal for that either. In spite of what you see in the movies, most women do not have orgasms from penile penetration.

---

[4] R.J. Levin, R.J., "The clitoris—an appraisal of its reproductive function during the fertile years: Why was it, and still Is, overlooked in accounts of female sexual arousal," Clin Anat 33 (2020): 136–145.

[5] W.H. Masters and V.E. Johnson, "Human Sexual Response." (1966) Toronto; New York: Bantam Books.

Some do, but it appears to be connected to indirect stimulation of the clitoris or other sensitive areas in the labia and along the upper wall of the vagina.

A more reasonable possibility is that female orgasm was important to favor pair bonding. Remember that as head size increased in our *Homo* ancestors, infants were born earlier in their development and required much more parental care. Not only was monogamy favored, but also the survival of offspring depended on the parents having a stable relationship. As monogamy was favored by selection to ensure the survival of infants, female orgasm could have played an important role to strengthen the bond between two individuals. Plus, achieving orgasm usually requires direct stimulation of the clitoris, so perhaps females of our ancestors used this as a way to assess a potential partner as a good caregiver and provider. A hormone called oxytocin—sometimes called "the brain's love drug"—is released during orgasm as well as childbirth and breastfeeding. It has the effect of making a woman feel tenderness and security, so it makes sense that female orgasm was important for making ancestral women feel safe and comfortable with a partner.

It looks like our evolutionary trajectory was different than bonobos; the purpose of the clitoris and female orgasm in humans was to improve pair bonding in our *Homo* ancestors. This was possible because the common ancestor of chimpanzees, bonobos, and humans was capable of female orgasm. But female orgasm and the associated clitoral morphologies were independently modified for pair bonding as brain size increased in our *Homo* ancestors and for maintaining social structures in bonobos. Since australopithecines had much smaller skulls and were most likely polygynous, the clitoris may not have served any function and may have begun to become vestigial, which would explain its small size in modern humans. But this trend was apparently reversed as head size increased in our *Homo* ancestors and as infants required increasing amounts of parental care. Consequently, monogamy and pair bonding were favored, resulting in a complex and highly functional clitoris. Female orgasm as a mechanism to favor pair bonding and monogamy in our *Homo* ancestors is the only explanation that makes sense from an evolutionary perspective—it is the only explanation that could have increased the fitness of our hominin ancestors.

This idea is somewhat speculative, but we got here the same way we came to conclusions about other human characteristics—by following the path that is best supported by the evidence and best guided by evolutionary logic. We understand that the human clitoris is not vestigial because of its complex structure. Furthermore, the connection between female orgasm and hormone stimulation is strong evidence that the clitoris was maintained and elaborated by selection as infants required increasing amounts of parental care and as monogamy and participation of both partners in caregiving were favored. Of all the ideas we have considered for female orgasm in humans, its role for pair bonding to improve the stability of monogamous relationships, and hence the survival of offspring, is the only one that is directly tied to the fitness of our ancestors.

*Summary*—We started this chapter discussing examples of vestigial structures in other animals, such as hip bones in whales. We then discussed some vestigial structures in humans, including arrector pili muscles, which are responsible for goosebumps, and some small muscles in our cartilaginous ears, which were important for ear movement in our early primate ancestors. While many previous authors have considered the clitoris as vestigial and a side product of female genital development, they did not recognize that the complexity of this organ says otherwise. We discussed how the clitoris is one of the most complex organs in the human body and has a higher density of nerve endings than the penis. We then examined the possible hypotheses that could explain the maintenance and elaboration of the clitoris and how female orgasm could have increased the fitness of our ancestors. We decided that it was unlikely that our ancestors had a social structure similar to bonobos; since frequent sexual interactions are not important for the social structure for two of the three living hominin species (i.e., it is present in bonobos but not chimpanzees and humans), it is unlikely to have been present in the common ancestor of all three species. As head size increased and stable monogamous relationships became important to ensure the survival of infants in our *Homo* ancestors, selection favored a more sensitive clitoris. The most likely explanation for the elaboration of the clitoris in our *Homo* ancestors is that female orgasm was important for pair bonding. It may have provided a mechanism for females to assess the quality of their potential partners. More importantly, the release of oxytocin during orgasm induced a feeling of tenderness and security and strengthened the bond and the stability of monogamous relationships.

# 9

# A mixing of peoples

*Launua's clan flourished in their new home. Sometime after her second child was born, she had a third, which did not survive past their first season, and then a fourth, who was healthy and strong. There was plenty of food and the clan grew in size, with their camps spreading up and down the valley. Occasionally other clans passed through the valley; some stayed, while others moved on. Now that she had a family, Launua found that her duties were mostly in the camp, helping to prepare food and looking after the numerous children. But she still made time to help with the gathering of roots because it was her favorite thing to do. Running across the savanna, she felt as free as she did when she was a child on her first forays. She loved the wide-open spaces and the discoveries she made, such as a new ravine to shelter from the sun or a new place to find roots for digging. She enjoyed crossing paths with gatherers and hunters from other clans. Some of the new settlers spoke strange words. She had to communicate using*

*Looking Down the Tree*. Mitchell B. Cruzan, Oxford University Press. © Oxford University Press (2025).
DOI: 10.1093/9780197805190.003.0009

*hand signs, but she always enjoyed the exchanges, and sometimes they would dig together. She began to notice that she could tell which clan someone belonged to by how they looked—mostly in their face. She recognized members of her own clan from different camps, and they seemed to have similar looks to members of her own camp. But people from some of the new clans looked very different, and some seemed to be much less skilled at digging or making tools, in spite of Launua's efforts to help them learn the best techniques. It was as if members of her own clan were always the winners; she could fill her basket with roots faster than any of the people from the other clans, and her clan's hunters always seemed to provide a good supply of meat, while the others struggled to find game. But the clans were not completely isolated from each other; a chance meeting sometimes led to mating, and occasionally desperate stragglers from other clans approached their camp begging for refuge. They were always accepted if they were women or children, but the clan was reluctant to accept wandering males— fearful that they may be undesirables that had been banished from another clan. Launua felt good about her clan and the other clans in the valley and was happy that everyone seemed to be getting along without any conflicts.*

Humans live in every habitable spot on the planet, as well as some places that are not so hospitable. There are even people who live year round in research stations in Antarctica, so humans have colonized every continent. But where did they all come from? How can we trace our ancestors back to their origins? We know a lot about the ancestry of many people alive today, especially families for whom we have a lot of information on their pedigrees, such as the royal family of the United Kingdom. But you don't have to be a member of the royal family to learn about your ancestry and uncover your genealogy, because there are a lot of tools and information available. You can have your DNA analyzed to tell you what fractions of your heritage come from different parts of the world. When portions of your DNA come from different geographic regions, it just tells you that at some time in the past one of your ancestors had a parent who hailed from that region and which side of your family that portion of your ancestry came from; it does not tell you which ancestor it was or when they lived there or give you specific information about your family tree. Most of us know who our grandparents are, but few know anything about our great-grandparents or great-great-grandparents. If you do extensive research on your family tree, you can probably find information going back several generations, but you will reach a point where no further information is available. You can't have descendants without ancestors, but how many generations back can we trace our ancestry?

If we keep going back generation by generation, then we should ultimately get to the beginning of our species, right? But was that somebody like Adam and Eve? And where did they come from? We don't have much information about the people who lived many generations in the past, and you typically can't find any real evidence of the earlier ancestors in your family tree—no photographs or even any paintings or drawings of what they looked like. But we know that they must have existed, so it's logical to conclude that if we trace back through time, we will eventually reach the ancestor of every human alive today. But it doesn't stop there. The earliest humans must have had ancestors too—ancestors that we share with other species. If we keep following the genealogy back even further, pretty soon we have a phylogeny of closely related species, including the great apes, then the primates, and then all of the mammals, and ultimately, we will have a phylogeny of all life on Earth, just as we discussed in an earlier chapter.

One of the reasons we often view ourselves as unique and superior to other species is that all of our closest relatives are extinct, and we are the only nonchimpanzee hominin species left. At one point in time there were at least half a dozen different hominins living in Africa and nearly as many in Eurasia, as indicated in Figure 9.1 as branches that end before the top of the frame. Imagine what it would have been like to meet one of them; would you try to make friends or run away? We are one of the only species alive today that does not have any close living relatives, which is one of the reasons it is difficult for us to understand our connections to other species. There is a bigger difference between us and other animals because there are no living intermediates between us and other great apes. We look very different from other animals—even from our closest living relatives, the chimpanzees and bonobos. And that makes it easier for us to accept the idea that we are special, that we are superior to other animals, or even that we were created separately from other animals by a supreme being.

But science tells a different story, one in which we share many characteristics with all other species—from our cellular chemistry to the form of our bodies— which tells us that our species evolved from common ancestors with other great apes and that we share ancestors with all life on Earth.

We know we share ancestry with other animals from studies of DNA, but there is a simpler way to see relationships among species. Many other species live with very close relatives—species that can interbreed to form hybrids. This is common because speciation is a slow process; as species diverge, there is a long period of time when they remain interfertile—they can still mate and produce viable hybrid offspring. We see evidence of the slow speciation process in many

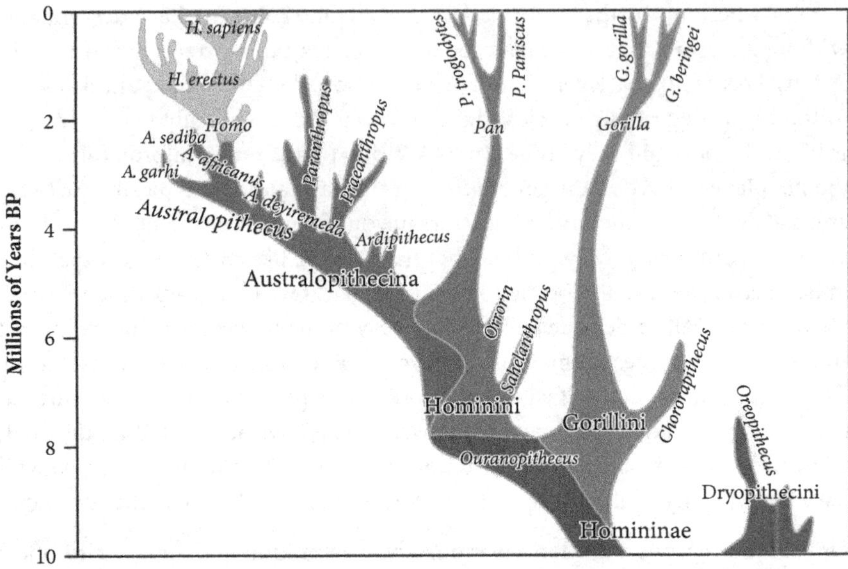


**Figure 9.1**  Historical diversification of the great apes including hominins and gorillas over the past ten million years. Living species are represented as branches that reach the top of the frame. Reproduced from Dbachmann (2017) via Wikimedia Commons, Creative Commons Attribution-Share Alike 4.0 International (CC-BY-SA-4.0).

animals you are familiar with. For example, wolves, coyotes, and domestic dogs are all interfertile and can hybridize. This is because they all share a recent common ancestor, and most recently, domestic dogs originated from gray wolves, so they can still hybridize. But even more distant relatives can interbreed, so coyotes are interfertile with both wolves and domestic dogs. The endangered red wolf from eastern North America has hybridized extensively with wolves and coyotes, and only a few individuals with pure red wolf ancestry persist in the wild (Figure 9.2).[1] These multiple interfertile species that can hybridize to varying degrees are referred to as *syngameons* and are widespread across the plant and animal kingdoms.[2]

Species in the family Equidae are another example of syngameons; horses are interfertile with donkeys, but they are not as closely related as wolves are with dogs and coyotes, so mules are sterile. Horses and donkeys can interbreed


[1] B. M. Von Holdt et al., "Persistence and Expansion of Cryptic Endangered Red Wolf Genomic Ancestry along the American Gulf Coast," *Molecular Ecology* 31 (2022): 5440–54.
[2] R. Buck and L. Flores-Rentería, "The Syngameon Enigma," *Plants* 11, no. 7 (2022): 895.

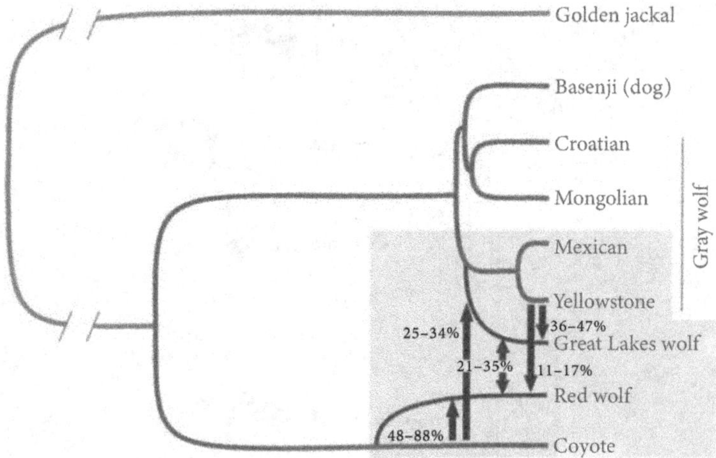


**Figure 9.2**  History of natural hybridization in the wolf-dog-coyote syngameon. Arrows between branches of the phylogeny indicate historical gene flow (hybridization) events for red wolves with coyotes and varieties of gray wolves. Reproduced with permission from B. M. von Holdt et al., "Whole-genome sequence analysis shows that two endemic species of North American wolf are admixtures of the coyote and gray wolf," Science Advances 2 (2016);e1501714. https://doi.org/10.1126/sciadv.1501714.

with zebras to produce different types of zebroids. Hybrids from some combinations are fertile, but other zebra hybrids are infertile. And then of course there is Napoleon Dynamite's[3] favorite animal, the liger, which is a hybrid between a male lion and a female tiger. Whether or not different species interbreed sometimes depends on where they live, so ligers are never seen in the wild, only in zoos where lions and tigers have been kept in the same enclosure. The genetic differences between lions and tigers are not as great as they are between horses and donkeys, so female ligers are fertile, while males are sterile. All of these cases, and many more, involve plants and animals that have diverged enough to become recognized as species but remain interfertile to some degree.

You might be wondering whether any two species can hybridize—maybe even a cat and a dog—but no, this is not the case. If we look, for example, at the phylogeny of species in the dog family in Figure 9.3, we see that those that are closely related, like gray wolves, domestic dogs, coyotes, and several of the African wolves and jackals, can all interbreed, but species that are more distantly

---

[3] https://en.wikipedia.org/wiki/Napoleon_Dynamite.

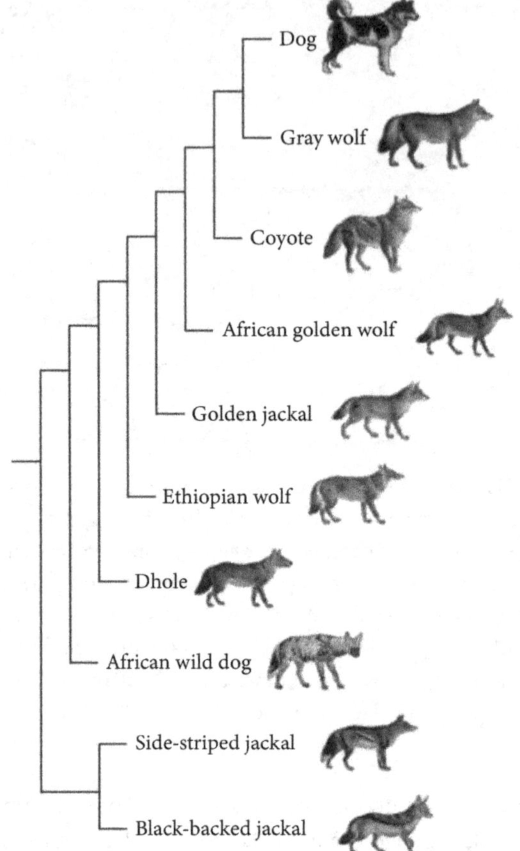


**Figure 9.3** Molecular phylogeny of Canidae including dogs, gray wolves, and their close relatives. Reproduced from Mivart, St. George Jackson (2017) via Wikimedia Commons.

related, like the African wild dog and various species of fox, can't interbreed with coyotes or wolves. This is true because the longer the divergence time for two species from a common ancestor, the more genetic differences they accumulate through mutations, and the less likely they are to produce viable hybrids.

But how does this process start—how do we get new species? It all starts with, a West and an East Coast subspecies. These two subspecies are effectively isolated from each other because the western one overwinters on the California coast, and the eastern one overwinters in mountainous regions of Mexico. There are many tropical butterflies that vary dramatically in their wing coloration patterns depending on where they live, and many of these have been designated as separate subspecies. In fact, any time a species is widespread, you are likely to find different varieties in different parts of its geographic range. This is true of widespread bird species. If you peruse bird identification manuals, you will find many species descriptions that include varieties and subspecies. The Ornithological Union meets annually to review genetic evidence for the classification of bird species and subspecies. In some cases, such as the Western Grebe, they decided that what was previously thought to be just a variety is actually a different species—the Clark's Grebe. In other cases, they discover that interbreeding and hybridization between species are extensive enough that they need to be grouped into one species. That was the fate of the red-shafted flicker, which is native to western North America, and the yellow-shafted flicker, which is native to the east; these two former species are now considered to be subspecies of the northern flicker.

In all of these cases, members of different subspecies are isolated from each other because they live in different places. The lack of genetic exchange allows them to diverge and respond to local selection. Geographic isolation is important for diversification and speciation as it allows isolated populations to become locally adapted and to follow their own evolutionary paths. During early stages of speciation, different varieties and subspecies within a species can hybridize, but if they continue to be separated, they will continue to accumulate genetic differences and will become less interfertile. Over longer periods of time, these lineages that started out as varieties and subspecies within a species will eventually become less and less able to interbreed and will eventually be considered as separate species.

Diversification into different subspecies and species also happened by the same processes described above during the history of our ancestors and their close relatives. *Homo erectus* was the first hominin to migrate extensively out of Africa to occupy large regions of Eurasia. Even though they had much smaoller brains than modern humans, *H. erectus* managed to cross seas to colonize southern Europe and islands in Indonesia. Such trips would have required months of planning and the construction of seafaring boats or rafts, so this species was clearly much more like modern humans than other great apes. There was apparently little movement among the regions after they were colonized. Over tens of thousands of years, isolated populations of *H. erectus* became morphologically distinct from each other. The fossils found in each region across Africa and Eurasia represent at least twenty-five different subspecies of *H. erectus* as well as some descendant species. These include *Homo heidelbergensis*, *Homo antecessor*, *Homo luzonensis*, *Homo neanderthalensis*, *Homo longi*, Denisovans (considered to be *H. denisova* or *H. altaiensis*, but possibly included in *H. longi*), and the diminutive *Homo floresiensis*—sometimes inaccurately referred to as "hobbits"—from the island of Flores.

We know much about early hominin species from fossil evidence. Fossils can tell us a lot, but they do not tell us the complete story of how early hominin species are connected with species that lived later. Fossil evidence suggests that *H. erectus* in Europe was possibly replaced by another hominin called *H. heidelbergensis*, which is the first hominin known to hunt large animals. This species was most likely a descendent of an early *H. erectus* subspecies in eastern Africa, but we can't know for sure because we do not have DNA from any subspecies of *H. erectus*. We know a lot more about the origin of other hominins from *H. heidelbergensis* because we have DNA obtained from this species and some of its descendants, including Neanderthals, Denisovans, and modern humans—*Homo sapiens*.

Fossil evidence can tell us a lot about when and where each of the different species of *Homo* lived, what types of tool-making technologies they had, and many things about their culture including art from cave paintings and burial rituals. It's more difficult to use fossils to provide a detailed picture of the time and place of the origin of each species and the areas they migrated to. For example, the earliest fossils of *H. erectus* from Africa are from 2 million years ago,[4] yet there is fossil evidence that *H. erectus* was already in eastern Asia around 2.1 million years ago.[5] This does not mean that *H. erectus* originated in Asia and then migrated back to Africa; it just means that there are too many fossils missing for us to put together the whole story. The problem is that fossilization of bones and teeth only occurs under a limited range of conditions. Then, the fossils have to be lucky enough to survive geological processes that might destroy them, and we have to be lucky enough to find them in good enough condition to identify which species they represent.

There's wide agreement that *H. erectus* originated in East or Southeast Africa as a descendent of an australopithecine species such as *Australopithecus afarensis* (perhaps via *H. habilis* as an intermediate species), so the first individuals of *H. erectus* must have pre-dated fossils from eastern Asia by a substantial amount of time. Similarly, *H. heidelbergensis* is only known from a jaw bone found in Germany and from a few specimens found in Spain, China, and Zambia, but it's assumed to have originated in eastern Africa.[6] Like most of the record of human history, the fossil evidence from the time around the middle of the Pleistocene

[4] J. G. Fleagle et al., eds., *Out of Africa I: The First Hominin Colonization of Eurasia*, Vertebrate Paleobiology and Paleoanthropology (Springer, 2010).

[5] Z. Zhu et al., "Hominin Occupation of the Chinese Loess Plateau since About 2.1 Million Years Ago," *Nature* 559 (2018): 608–12.

[6] L. T. Buck and C. B. Stringer, "*Homo heidelbergensis*," *Current Biology* 24 (2014): R214–15.

is sparse. The problem is worsened by the fact that hominins were moving around quite a bit. They migrated as far as eastern and northern Asia and northern Europe over a relatively short period of time. There was also quite a lot of interbreeding, which further obscured distinctions among different species and subspecies. Consequently, there is confusion concerning the identification of species and subspecies during this time, so human history during the middle Pleistocene is sometimes referred to as the "muddle in the middle." This has led to disagreements among paleontologists about the identification of different species. For example, *H. heidelbergensis* has been considered by some as a subspecies of *H. erectus*[7] (referred to as *Homo erectus* ssp. *heidelbergensis*), as have several other species of *Homo*. The wide and apparently continuous variation within and among species and subspecies during this time was probably due to interbreeding among species and *chronological speciation*, the gradual change from one species or subspecies to the next over time. For example, chronological speciation occurred as *Homo hablis* descended from *Australopithecus afarensis* or a similar species, and as a branch of that species evolved into *H. erectus* (or *H. erectus* descended directly from an australopithecine species). Then the many subspecies of *H. erectus* diversified into several hominin species across Asia and Africa, including *H. sapiens* in eastern Africa.

Where fossils provide only a minimal understanding of historical migration patterns, evidence from DNA of living individuals can provide much more detailed information about when our ancestors colonized different parts of the world. As we discussed previously, we can compare DNA sequence data among species to estimate a phylogeny of their relationships and then calibrate the nodes of the phylogeny (where branches join) with fossils to estimate the timing of divergence from common ancestors. We can also examine DNA in to evaluate the hybridization history of our human ancestors with other species including Neanderthals, Denisovans, and at least one other unknown archaic hominin. All of this information is important for us to understand our history and origins from gene genealogies, which can reveal the paths our ancestors took as they migrated out of eastern African and when they arrived in different geographic regions around the world.

As we discussed earlier, the DNA code is the genetic material that is passed from one generation to the next, and it provides accurate estimates of relatedness and a reliable mechanism for understanding the evolutionary history of species. The large amount of information available from genome sequences means that relationships can be determined for both closely and distantly related

---

[7] K. Harvati, "100 Years of *Homo heidelbergensis*—Life and Times of a Controversial Taxon," *Mitteilungen der Gesellschaft für Urgeschichte* 16 (2007): 85.

individuals. You might remember from a previous chapter that the evidence for relatedness based on DNA—based on the molecular clock—is substantial; relationships among different species can be estimated using advanced mathematical methods and do not require guesswork the way fossils do. DNA provides us with reliable information about how people from different parts of the world are related, so let's take a look at how we can use DNA to understand the origin of modern humans.

As we discussed in a previous chapter, we can use gene genealogies to understand relationships among groups of individuals within a species. The most well-known human gene genealogy is based on the mitochondrial genome in humans.[8] Mitochondria are organelles within our cells that are descended from early bacterial symbionts of unicellular organisms that lived around two billion years ago. Mitochondria have their own genome, which is similar to the genomes of modern bacteria. They are symbiotic and serve important functions inside of our cells; we cannot live without them, and they cannot live without us.

If you are familiar with the Star Wars movies, you might remember the scene in *The Phantom Menace* where Qui-Gon explains to a young Anakin Skywalker about *midichlorians*. The name is a combination of two symbionts—mitochondria and chloroplasts—that live inside the cells of eukaryotes, which include all life on Earth except for archaea and bacteria. Mitochondria are commonly called "the powerhouse of the cell," and chloroplasts are responsible for the conversion of light energy to chemical energy, including sugars and carbohydrates, in plants. There are a lot of similarities between midichlorians and mitochondria. For example, both are present in all living things—everything except bacteria and archaea—and are inherited from parents to offspring. But one difference is that mitochondria are inherited from our mothers because they are carried by eggs, not by sperm. Another difference is that, in Star Wars, each person is born with a certain concentration of midichlorians, and it doesn't change, but you can increase the number of mitochondria in your muscle cells through exercise.

A genealogy based on mitochondria is really a genealogy of women because people only inherit mitochondria from their mothers, so mitochondria are passed down from one female generation to the next. It represents a family tree of all people alive today that is based on a smaller sample of individuals from around the world; it's like a phylogenetic tree within a species that is part of a larger phylogenetic tree showing relationships among species.

We can make a gene genealogy based on human mitochondrial DNA and from the Y chromosome the same way we build a family tree based on last names, but

---

[8] P. Endicott et al., "Evaluating the Mitochondrial Timescale of Human Evolution," *Trends in Ecology & Evolution* 24 (2009): 515–21.

with DNA, the lengths of the "names" are thousands of letters long. The other difference is that these DNA "names" change by one letter at a time with each new mutation—it's like the molecular clock we talked about earlier. Each time there is a new mutation, we get a new branch. If we go back up the tree, we can trace the name changes with each mutation until we get back to the original DNA sequence that existed many generations in the past. It's important to note that the ancestors of all the people alive today can be traced to a single genome of a hypothetical person that lived many generations in the past. As we said before, they are hypothetical because the genotype we estimate will depend on the sample of living humans that we start with. We call this genotype the *most recent common ancestor*—the MRCA. More commonly, she is known as the "Mitochondrial Eve," but to understand why this isn't a real person, we need to look at how this tree is made.

If you imagine a number of DNA sequences lined up with each one representing a single person, you will see that most of the letters—DNA nucleotides—are the same across all sequences. But sometimes they are different, with some sequences having one nucleotide at a particular position and others having a different nucleotide. You might note that at one position, most individuals have an A while a few have a T. We can make a good guess about what the MRCA sequence had at this position just by noting the frequency difference; there are more people with an A at this position than there are people with a T. The simplest explanation for this difference is that the original letter was A because it is more common in people today, and then the T arose by mutation some generations later, so it is present in fewer people. If we do this for all of the variable nucleotide positions, we can determine three things. The first is the most likely original sequence of the common ancestor of all of the people we took samples from—in this case it was most likely to be an A at this position. The second is, using what we know about how frequently mutations occur, an estimation of how long ago that hypothetical person lived by counting up all the mutations from all of the individuals sampled across the mitochondrial genome. The third thing is, as we discussed in a previous chapter, where people with DNA sequences most similar to the original live today—in eastern Africa. The mathematics and computer processing of DNA sequence data are a bit more complicated, but this simple example gives you a good idea of how we can use information from our DNA to learn things about our ancestors. When these analyses were done for a very large number of people from around the world, we learned from application of coalescence theory that the group of our ancestors representing the Mitochondrial Eve lived around 120,000 to 156,000 years ago.

As we discussed previously, the DNA sequence we estimate for the MRCA will change slightly depending on whose DNA are sampled. For example, we could take mitochondrial DNA samples from everybody in a particular town and use them to make a gene genealogy that would identify the hypothetical ancestor they all share. Even though the identity of that person is hypothetical, the idea that everyone in that town shares an ancestor is real. If we sampled people from a different town, we might come up with a slightly different sequence for the Mitochondrial Eve. When scientists develop gene genealogies, they make sure to sample a large number of people from around the world to get the best estimate of the MRCA's genotype. These genealogies tell us that we are all related to some degree or another. But gene genealogies can tell us much more than that; we can use them to trace the ancestry of every living human back to our origins in eastern Africa.

While we have learned much from the application of coalescence theory to variations in the mitochondrial genomes of living humans, much more depth and detail on the demographic history of our lineage can be attained using data from sequencing of the entire nuclear genome. Mitochondrial genomes are a mere 16,569 nucleotide pairs in length, and the Y chromosome consists of 59 million nucleotide pairs. These are miniscule compared to the entire human nuclear genome which includes all twenty-three chromosomes consisting of more than 3 billion nucleotide pairs. The multimagnitude increase in information we can obtain from the entire nuclear genome can provide substantially greater detail about the history of our lineage going back through time, including descriptions of demographic events such as historical changes in population size.

An analysis of the nuclear genomes from 1,000 individuals from Africa and across other parts of the world revealed a severe *population bottleneck*—a period of much-reduced population size—from nearly 100,000 individuals down to fewer than 1,280 between 930,000 and 813,000 years ago.[9] During this time our direct ancestors would have been considered to be a subspecies of *H. erectus*. This near-extinction event was caused by a Pleistocene glacial maximum that induced 100,000 years of severe drought in eastern Africa, which in turn caused changes in the composition of wildlife populations as several species went extinct. Population bottlenecks are often associated with rapid evolution, which is evident in the coalescence tree by time periods of higher densities of new mutations. It is thought that a speciation event occurred in our ancestral lineage during this time due to a chromosome rearrangement,[10] perhaps as the origin of our *H. heidelbergensis* ancestor from our ancestral subspecies of *H. erectus*.

[9] Wangjie Hu et al., "Genomic Inference of a Severe Human Bottleneck during the Early to Middle Pleistocene Transition," *Science* 381 (2023): 979–84.

[10] K. Poszewiecka et al., "Revised Time Estimation of the Ancestral Human Chromosome 2 Fusion," *BMC Genomics* 23 (2022): 616.

As we discussed previously, this severe bottleneck could represent the origin of increased cooperation and the inclusion of exclusively homosexual individuals to act as helpers within the clans of our ancestors, which occurred in response to the much higher amount of childcare required as a result of increasing brain size. The surviving populations would have all shared these characteristics, which would have contributed to their success as they expanded across the African continent and the rest of the world. The first migration occurred as *H. erectus* clans traveled out of Africa around 500,000 years ago, and then as *H. sapiens* migrated throughout Africa and around the world around 70,000 years ago. Monogamy, cooperation, and the inclusion of homosexuals within surviving clans became intrinsic to our species, so these characteristics are shared across all societies of humans around the world.

The wave of migration of *H. sapiens* out of Africa shown in Figure 9.4 resulted in the colonization of all of the other continents (except Antarctica) starting around 70,000 years ago, during the last glaciation of the Pleistocene.[11] Within Africa, artifacts associated with *H. sapiens* have been found as early as 300,000 years ago in Morocco, but the earliest fossil evidence of *anatomically modern humans* (AMH; sometimes called Cro-Magnons) comes from Ethiopia around 200,000 years ago.[12] As we said before, this does not mean that our species originated in northern Africa; gene genealogies clearly show eastern Africa as the epicenter of early migrations colonizing the different regions of the continent and a later wave of migration that resulted in the colonization of regions outside of Africa. There is evidence of much earlier migrations out of Africa in Europe and maybe even all the way to the west coast of North America, but these may have been small groups of early humans who died out before the next wave of human migrants arrived; they did not leave lasting effects on the genomes of native people in those regions.[13]

It's a little shocking to realize that fewer than 1,000 individuals migrated out of Africa 70,000 years ago to initiate the spread of humans around the world and to establish today's populations totaling more than 6.5 billion people. This migration originated from a population of nearly 30,000 individuals that had been maintained since the near-extinction event that occurred around 900,000 years ago. These estimates come from historical demographic analysis of DNA sequence variation using coalescence theory and refer to genetically distinct individuals living at that time. Since we know that people lived in inbred

---

[11] G. Stix, "The Migration History of Humans: DNA Study Traces Human Origins across the Continents," *Scientific American*, July 2008.

[12] E. M. L. Scerri et al., "Did Our Species Evolve in Subdivided Populations across Africa, and Why Does It Matter?," *Trends in Ecology & Evolution* 33 (2018): 582–94.

[13] S. Lopez et al., "Human Dispersal Out of Africa: A Lasting Debate," *Evolutionary Bioinformatics* 11, Suppl. 2 (2016): 57–68.

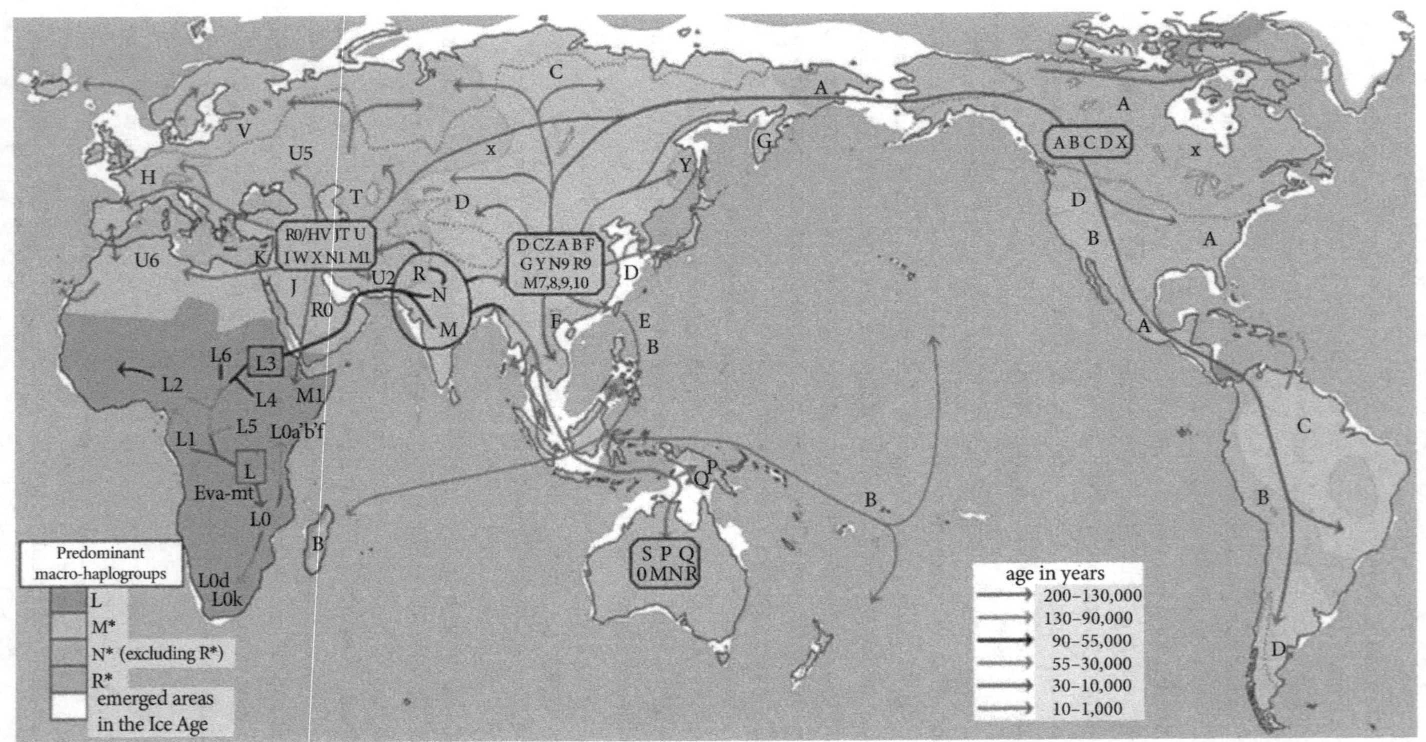


Figure 9.4  Human migratory patterns over the past 100,000 years based on variations in mitochondrial genomes. Letters indicate distinct mitochondrial genotypes. Reproduced from Maulucioni (2014) via Wikimedia Commons, Creative Commons Attribution-Share Alike 3.0 Unported (CC-BY-SA-3.0).

clans where the members were closely related, we can guess that the actual number of people participating in this migration out of Africa was probably greater; it simply means that there was enough genetic diversity across clans to be equivalent to about 1,000 genetically distinct individuals.

The consequences of this second population bottleneck are evident in the genetic diversity among humans around the world. Much earlier, there had been three major migrations from eastern Africa that colonized regions of southern, western, and northern Africa. Consequently, there is much more genetic diversity among people native to Africa than among people across the rest of the world. If we examined a gene genealogy of people across all continents, we would see three major branches within the African continent, each of which has many smaller stems emanating from eastern Africa. A fourth, much sparser branch with few stems represents the diversity across the remainder of the world. The "bushiness" of the three African branches reflects the much higher genetic diversity within these groups, compared to the group that was subject to a bottleneck during the migration out of Africa. Consequently, the large majority of genetic diversity in humans living today remains in Africa. This is evident as more rare genetic variants are present in Africa (64 percent) than any other continent.[14]

The process of colonization of regions around the world continued from 70,000 years ago and culminated within the last few thousand years. The first migration of modern humans out of Africa crossed the Red Sea into what are now the Arab countries (Figure 9.4). These clans followed along the southern coast of Asia and then headed south through Indonesia and into Australia by 65,000 years ago. During this time, the climate was much colder across Eurasia, and much of Europe and northwest Asia was covered by the Eurasian ice sheet. Sea levels were lower, so these human migrants were able to travel mostly overland in regions that are currently covered by water. This first route of human migration out of Africa was probably guided by the avoidance of colder climates farther north and the availability of food in coastal regions. Humans following this route encountered Denisovans but not Neanderthals, who occupied regions farther north in the Middle East, Europe, and western Asia. Consequently, the genomes of people native to the regions along this migration route, including Malaysians and Aboriginal Australians, show evidence of historical interbreeding with Denisovans but not with Neanderthals.[15]

Migration into colder regions of Eurasia proceeded at a slower pace with migrations from the southern coast of Asia extending into the Middle East,

[14] M. C. Campbell and S. A. Tishkoff, "African Genetic Diversity: Implications for Human Demographic History, Modern Human Origins, and Complex Disease Mapping," *Annual Review of Genomics and Human Genetics* 9 (2008): 403–33.

[15] D. Reich et al., "Denisova Admixture and the First Modern Human Dispersals into Southeast Asia and Oceania," *American Journal of Human Genetics* 89 (2011): 516–28.

southern Europe, and China between 45,000 and 50,000 years ago. These migrations were probably motivated by food availability as they searched for game, such as the large mammals of the Pleistocene megafauna. Humans expanding into more northern regions encountered Neanderthals, Denisovans, and other archaic hominins and may have competed with them for hunting territories, which slowed the progress of human migration across Europe. Humans migrating from equatorial regions would have lacked the adaptations and skills for survival in colder climates but obtained them through cultural exchange and interbreeding with Neanderthals and Denisovans.

Migrations and colonization of other regions were both slowed and facilitated by the glacial period that began around 110,000 years ago and ended about 10,000 years ago. The low sea levels allowed humans from eastern Asia to cross the Bering Land Bridge that connected Asia and Alaska around 25,000 years ago. There are competing ideas about how their southward migration in the Americas proceeded. One is that they followed megafauna game through North America along corridors between the massive ice sheets. The other is that the earliest humans to colonize North and South America were primarily foragers who followed the coastline on ice-free land along the shore or by watercraft. There is evidence of early settlements in Central and South America that supports this idea, but sea levels were more than a hundred feet lower at that time, so traces of the passage of humans along the coast may currently be underwater. Migrations into southern regions of North America didn't occur until 16,000 years ago, and into South America around 14,000 years ago. Similarly, northern regions of Europe and Asia were not colonized until around 12,000 years ago. Eastern Canada and Greenland were colonized within the last 5,000 years.

Major seafaring migrations began within the last 3,000 years.[16] Humans living in what is now known as Taiwan developed advanced sailing technologies such as catamarans, outrigger canoes, and *crab claw sails* (a fore-and-aft triangular sail with spars along the upper and lower edges). They spread eastward to colonize the Southeast Asia Islands, Micronesia, Melanesia, and Polynesia, eventually reaching the Hawaiian Islands and New Zealand within the last 1,000 years. Migration westward from Taiwan resulted in colonization of Madagascar by 2,000 years ago.

From the discussion above, we can see that every person on this planet can trace their origins back to eastern Africa, where our species descended from a subspecies of *H. erectus* (possibly with *H. heidelbergensis* as an intermediate),

---

[16] A. G. Ioannidis et al., "Paths and Timings of the Peopling of Polynesia Inferred from Genomic Networks," *Nature* 597 (2021): 522–26.

which had descended from an earlier australopithecine species. The fossil evidence of all the stages of our history is incomplete, but molecular phylogenies and genealogies tell an indisputable story: we share common ancestors with every other person living today or in the past, with the other great apes and primates, and with all species on Earth. Science doesn't provide us with any morals or answers to ethical questions of right or wrong or tell us how to treat each other—that's the role of cultural morality guidelines, philosophy, and religion. But it does tell us that regardless of our skin color, nationality, political affiliation, or cultural background, we are all related to each other; we are all members of the same human family. Science just gives us the facts, and it's up to you to decide how to use that information in your daily lives.

*Summary*—We started this chapter discussing how we can use fossils and DNA to understand relationships among species. We emphasized that speciation is a slow process; as pairs of species diverge from common ancestors, there is often the potential to hybridize, but more distantly related species are reproductively isolated due to their accumulation of unique genetic mutations. We discussed the process of speciation; from geographic variation within species to genetic divergence on independent evolutionary paths, eventually enough genetic differences accumulate for reproductive isolation. We discussed *H. erectus* as a species that spread out of Africa to occupy regions across Eurasia. These isolated populations diverged to become different subspecies and sometimes became different enough to be considered a new species. *H. erectus* represents the first species migrating from eastern Africa. They were followed by *H. heidelbergensis* and later *H. sapiens*, both of which originated from the same region in eastern Africa. We learned that our ancestral lineage suffered a near-extinction event around 900,000 years ago due to severe drought in eastern Africa, which resulted in a chromosomal rearrangement and the possible origin of *H. heidelbergensis* leading up to the origin of *H. sapiens*.

We discussed how *H. sapiens* originated in eastern Africa, probably from a subspecies of *H. erectus* or from *H. heidelbergensis*, and spread in three major migrations to occupy southern, western, and northern Africa. A fourth branch of humans migrated across the Red Sea along the southern coast of Asia and into Australia. This branch of humanity has much lower genetic diversity due to the population bottleneck that occurred early during their travels from Africa. All populations of humans outside of Africa are derived from this small group of migrants who traveled out of Africa around 70,000 years ago.

Clans of *H. sapiens* first crossed the Red Sea and spread southward along the coast through what is now Malaysia and then into Australia. Along the way they encountered Denisovans but not Neanderthals. Other clans spread eastward into eastern Asia, and some crossed the Bering Land Bridge to enter North America and eventually South America. Spread of humans farther northward in Europe occurred at a slower pace because much of the area was still occupied by Pleistocene ice sheets. These ancestors faced challenges because they had to compete with Neanderthals, Denisovans, and possibly other archaic hominins, and they were poorly adapted for life in colder climates. Interactions with other hominins resulted in cultural exchange and interbreeding, which provided the adaptive genetic variation necessary for the continued success of our ancestors. Colonization of the Pacific Islands and Madagascar occurred over the last few thousand years as occupants of Southeast Asia developed more advanced sailing technologies.

# 10
# The strangers

*The valley that was once so wonderful began to feel less and less like a good place. As local sources of food and firewood became depleted, members of Launua's clan had to venture farther to find them, and conflicts began to break out among the separate camps. The members of Launua's camp became afraid, and more than once they were chased away from good places to dig for roots by people from other camps. They began to think of themselves as a separate clan, and finally the decision was made to leave on their own without the other camps. Launua's oldest children were able to walk on their own and even helped carry supplies, but she had one child who was nursing and a second who was a toddler. The youngest children were carried or rode on the sleds with the food, water, and other supplies. They traveled for many days and gathered food or replenished their water supply whenever they could, but they did not find a river or lake large enough to serve as a source of water during the dry season. As they traveled, the land became very dry, and places to find water were few,*

*Looking Down the Tree*. Mitchell B. Cruzan, Oxford University Press. © Oxford University Press (2025).
DOI: 10.1093/9780197805190.003.0010

*until they could not find any at all. Their supplies of food and water dwindled, but they pressed on, because it was the only thing they could do. Finally, the scouts came back to the group with a report of a very large lake. Several of them ran ahead to get the first drinks, but as the group got closer, they could see the others running back from the shore. The water tasted very bad, and some who had drunk the water were retching and looked ill. Launua offered one of them the last of her water, and they were grateful. The clan did not know what to do, but one of the scouts spotted smoke from campfires farther up the shore. Hoping there would be good water there, they set out in the direction of the camp.*

*As the group approached the camp, the scouts came running back. They were frightened by something, and the clan soon understood why. Walking toward them were the strangest people they had ever seen. Their skin was the color of bones, and the long hair that flowed down their backs was the color of dried grass. The faces of the men were covered with thick curly hair that was so long it hung across their chests. They were larger—not taller, but with big chests and thick arms and legs—and they were making noises that Launua's clan did not understand. They stopped with some distance between the groups, and Launua thought that members of the other group also looked frightened. Finally, Launua's partner—now a senior member of the clan—took some tentative steps toward the strangers, and one of their women matched his steps. As Launua's partner approached the stranger, he glanced back for reassurance. Not looking where he was stepping, a rock rolled under his foot; he fell and landed on his behind. Launua cried in fear to see the stranger standing over him with a spear, but the stranger suddenly started to laugh. She reached out a hand and helped Launua's partner to his feet. Launua's partner started to laugh, and then everybody was laughing.*

*The two clans started to mingle. Now, they were more curious about each other; they approached tentatively, but they were much less afraid of these strange people. Launua approached one of the women. She reached out to touch her skin and was surprised to see that it was not covered with paint or dust—it really was a light color. She marveled at the contrast with her own coal-black skin. She touched the women's pale, straight hair, and the stranger was equally curious about Launua's bushy hair, which was dark like the color of burnt wood. Launua noticed a string around the stranger's neck, and she reached out to touch it. There were curved flat stones with a design on the surface and holes for the string to pass through. Launua thought it was the most beautiful thing she had ever seen, and the strange woman smiled as she pulled it over her head and placed it around Launua's neck. She smiled back and ran to show her partner the beautiful thing the stranger had given her.*

*Launua's clan camped near the strangers for several days. There was good water from a stream that flowed into the lake, and the strangers shared their food with them. Some of the food came from the large lake with bad water, and they were surprised that it tasted very good. The strange men were curious about the spears that Launua's clan carried, which had sharp stone spearheads while the strangers' spears were wooden sticks with sharpened tips. Launua's people gave them some of the stones they used to make spearheads and tools and taught them the technique to strike the stone just right to make a sharp point or cutting tool. Launua learned that the thin, curved stones she'd seen as necklaces and other decorations came from animals that lived in the lake. Two curved stones were pressed together to make a hollow space and the animal lived inside. They were good to eat and easy to find but had to be cut from the rocks on the shore with a sharp stone. The strange women showed her how to use a stick to bore holes in the stones so she could make her own neck string. It was often difficult to communicate because neither of the groups understood the sounds of the other, but they managed quite well with hand gestures.*

*Launua's clan learned from the strangers that there was another place with good water a few days' walk away along the shore. As they gestured their thanks and prepared to set off, one of the young men from Launua's clan wanted to stay. He had mated with one of their women and the two of them stood together with the stranger clan. There was some discussion and concern; Launua's clan was small, and they could not afford to lose a member. The people in the stranger clan started to look concerned; they did not want a conflict. But then, one of the young women from the stranger clan walked over to one of the men in Launua's clan and grasped his hand. Both of them looked pleased and the clans decided it was a fair trade; the stranger woman would come with Launua's clan and the young man would stay with their clan.*

*After several days of walking along the shore, the clan came to a lush valley with plenty of water and good places to find food. They camped and decided that they would stay. They used the skills they had learned from the stranger clan to find food in the large lake and from the unfamiliar plants that grew in the region. Launua was happy to see her clan so content, and everyone was pleased to see that several of the women—including the bone-skinned stranger—were growing round bellies and soon would have new babies.*

In the last chapter we discussed two questions that could be answered using gene genealogies: how long ago and where did our most recent common ancestors (MRCAs) live? As we discussed previously, we know from the fossil evidence that humans inhabited eastern Africa around the same time as our MRCAs did based on estimates from the gene genealogy, and people living in this region today have mitochondrial DNA sequences that are the closest match to the Mitochondrial Eve. Note that she represents a group of individuals that are the MRCAs of all living humans. At the time that Mitochondrial Eve lived, many thousands of women were alive—Eve was not alone in the world with Adam. By chance, all the other maternal lineages died out, so only one can be traced back to our MRCAs—the hypothetical Mitochondrial Eve. But these were not the first humans. As we mentioned before, fossils of *Homo sapiens* and artifacts associated with them have been dated as far back as 200,000 to 300,000 years ago.[1] It was around 300,000 years ago that more finely crafted tools appeared, such as throwing spears with flaked stone points;[2] earlier thrusting spears were basically just sharpened sticks and had been used for the previous 1.5 million years. This represented a major advancement that improved hunting success and ultimately facilitated the spread of humans out of Africa. Multiple small migrations of humans out of Africa occurred over the last 200,000 years, but these did not persist. The major migration event around 70,000 years ago established all populations of humans around the world outside of Africa.

We discussed how modern humans were not the first hominins to migrate out of Africa. Remember that *Homo erectus* colonized large regions in Eurasia staring around 500,000 years ago following the near-extinction event of our ancestral lineage in eastern Africa, and later, they may have been replaced by *Homo heidelbergensis*. This species of *H. erectus* later diversified into Neanderthals and Denisovans, and perhaps other archaic hominins. Eventually these other species

---

[1] E. M. L. Scerri et al., "Did Our Species Evolve in Subdivided Populations across Africa, and Why Does It Matter?," *Trends in Ecology & Evolution* 33 (2018): 582–94.

[2] Y. Sahle et al., "Earliest Stone-Tipped Projectiles from the Ethiopian Rift Date to >279,000 Years Ago," *PLoS One* 8 (2013): e78092.

disappeared, but only after a long period of time when our direct ancestors lived alongside them in Africa and later in Eurasia. The questions we want to ask now are, what happened to the other hominin species, and what impact did they have on the evolution of our species?

From comparisons of DNA, we know that *H. heidelbergensis* is closely related to Neanderthals in Europe and western Asia, to Denisovans in northern and eastern Asia, and probably to at least one other hominin in Asia. Note that we do not know how *H. erectus* fits into this picture because we do not have DNA from any of its subspecies. All these species were closely related, so they could have easily interbred with each other. DNA evidence tells us that Neanderthals and Denisovans interbred with each other and with at least one other unknown hominin living in Eurasia. Once modern humans migrated out of Africa and spread across Europe and Asia, they interbred with all of these species.

You probably know that, depending on where your ancestors come from, you may have a small amount of Neanderthal and/or Denisovan DNA in your genome. If we look across all humans alive today, we can find about 20 percent of the Neanderthal genome, but any individual carries only about 2 percent of DNA that can be traced to Neanderthals.[3] Melanesian and aboriginal Australian genomes carry between 3 and 6 percent of Denisovan DNA. The only way this could have happened is if our ancestors interbred with Neanderthals and Denisovans to make hybrid offspring.

But how could such a small percentage of our genome come from Neanderthals and Denisovans? We get 50 percent of our genes from each of our parents, so how could we have only 2 percent from Neanderthals? The answer is that the 50 percent rule only applies to the first hybrid generation (Figure 10.1), which had an equal mixture of human and Neanderthal DNA. But then those hybrid individuals mated with humans, so their children had a smaller amount of Neanderthal DNA—about 25 percent. For example, the character B'Elanna Torres from the *Star Trek: Voyager* series was a first-generation Klingon-human hybrid, so she was 50 percent Klingon. She had a baby with the human character, Tom Paris, so their child was 25 percent Klingon. This mating process is called backcrossing and has been regularly used by plant and animal breeders to transfer desirable traits from one closely related species into another. As you can see in Figure 10.1, the proportion of one ancestral genome becomes smaller with each generation of backcrossing. In the same way, with each generation of backcrossing between hybrids and humans, their offspring carry smaller and smaller amounts of Neanderthal DNA. Mating between hybrids and humans was repeated generation after generation so that the Neanderthal DNA was

[3] L. Chen et al., "Identifying and Interpreting Apparent Neanderthal Ancestry in African Individuals," *Cell* 180 (2020): 677–87.

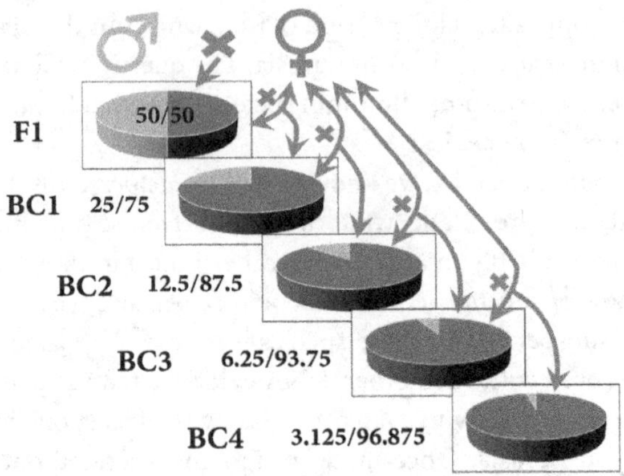



**Figure 10.1** Reduction in genomic representation due to hybridization and repeated backcrossing. Numbers indicate the proportional percentage of each genome for each hybrid generation. M.B. Cruzan (own work).

diluted in the human population. Today, many humans who are descendants of these hybrids have only a small percentage of their genomes from Neanderthals. This dilution of DNA occurs as portions of sister *chromosomes* are swapped back and forth via *recombination* when sperm and eggs are formed. After recombination, each chromosome is like a mosaic, with some stretches of DNA inherited from one ancestor and other stretches inherited from the other ancestor. Now many of us carry only small fragments of DNA as continuous regions of chromosomes that we inherited from Neanderthals. The same process also accounts for small amounts of Denisovan DNA in many human populations around the world.

Some might think that interbreeding between our ancestors and different archaic hominin species is of little or no consequence, whereas others might think that this somehow diluted the purity of our species. Some have taken the fact that European human ancestors interbred with Neanderthals and twisted it to infer that modern Europeans are somehow superior to other races.[4] Even though some characteristics in modern humans may have been acquired by interbreeding with other hominins, the contribution of Neanderthals and other archaic hominins to our genome is very small. Furthermore, native populations across Eurasia, northern Africa, and North and South America all

[4] L. Weasel, "How Neanderthals Became White: The Introgression of Race into Contemporary Human Evolutionary Genetics," *American Naturalist* 200 (2022): 129–39.

show evidence of hybridization with Neanderthals, so this is not a characteristic that is exclusive to white Caucasians of European descent. This is not a unique characteristic of our human history; interbreeding among closely related species is extremely common across the animal and plant kingdoms. As we mentioned before, the process of speciation is very slow, so there can be ample opportunity for closely related lineages to intermix before strong reproductive isolation develops. We have learned from numerous studies of plants and animals that hybridization often acts as an evolutionary stimulus because it allows divergent lineages to exchange the advantageous mutations they have acquired.[5] Hence, we should view these historical genetic exchanges as beneficial, and it's likely that the success of the ancestors of all living humans was improved by adaptations they acquired from these episodes of hybridization with other hominin species.

Remember from a previous chapter that the variations in our mitochondrial genomes allowed us to trace our ancestry back to eastern Africa. Since mitochondria are only inherited from our mothers, does that mean that only human females mated with Neanderthal and Denisovan men and not the other way around? That is one interpretation, but we can't know for sure. It could have been that Neanderthal and Denisovan mitochondrial DNA was present in ancestral human populations but was lost over many generations by chance or possibly by selection. But the thing we do know for sure is that interbreeding between humans and Neanderthals happened multiple times, and only for populations of humans that migrated out of Africa, although there was back-migration of Eurasian humans who had hybridized with Neanderthals into northern Africa. Human ancestors that stayed in middle and southern Africa never interbred with Neanderthals or Denisovans, so descendants who live in these regions today do not have DNA from those hominin species. Instead, there were other archaic hominins in Africa that the ancestors of modern Africans interbred with.

We can understand the contributions of archaic hominins to our evolutionary history if we look at Figure 10.2, which represents the genetic diversity of modern humans and includes the contributions of archaic hominins. The first thing you might notice is that one branch leads to all populations around the world except for Africa. This is the branch that represents the major migration out of Africa that occurred around 70,000 years ago and resulted in the colonization of all regions of the world except central and southern Africa. Connections from Denisovan and Neanderthal branches represent episodes of hybridization with those species. But, as mentioned above, the majority of genetic diversity in our

---

[5] S. B. Yakimowski and L. H. Rieseberg, "The Role of Homoploid Hybridization in Evolution: A Century of Studies Synthesizing Genetics and Ecology," *American Journal of Botany* 101 (2014): 1247–58.

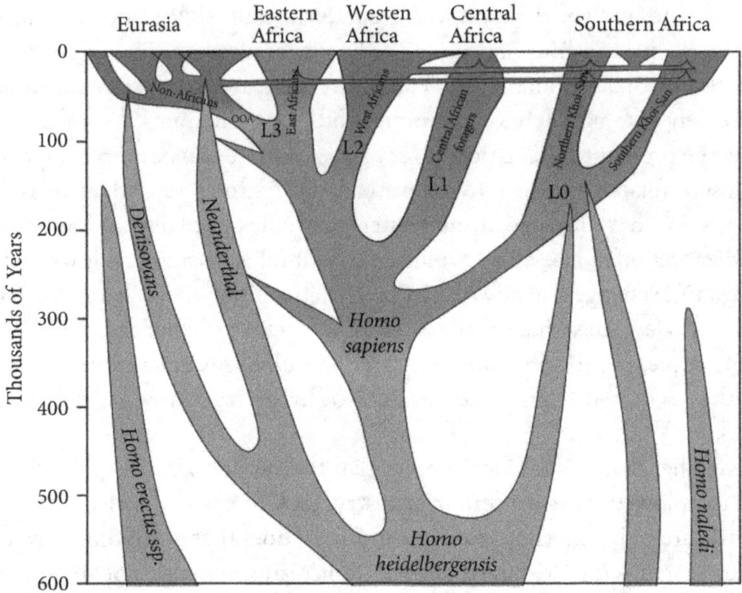


**Figure 10.2** Historical diversification of modern humans. Intersections between lineages indicate hybridization with archaic hominins.

species is found among Indigenous populations in different regions of Africa, which is represented by separate branches for eastern, western, central, and southern Africa. This high level of genetic diversity is consistent with the idea that all humans originated from Africa and that the inbred clans that migrated out of Africa carried only a small fraction of the genetic diversity that was present in Africa. We all have ancestors from Africa no matter where we are from, but some of us have much more recent African ancestors than others.

The spread of our human ancestors out of Africa some 70,000 years ago was relatively rapid—perhaps because they had developed some technological and cognitive advantages that were lacking in other hominins. Such an advantage would have appeared in African populations many generations before. In fact, there is some evidence of previous migrations into territories occupied by Neanderthals as early as 185,000 years ago, but these were limited and those settlements apparently did not persist.

The fact that there are no native populations in central or southern Africa with Neanderthal DNA tells us that the humans migrating out of Africa around 70,000 years ago hybridized with Neanderthals and Denisovans. We can see that there was a lot of interbreeding among hominins in Eurasia and Africa.

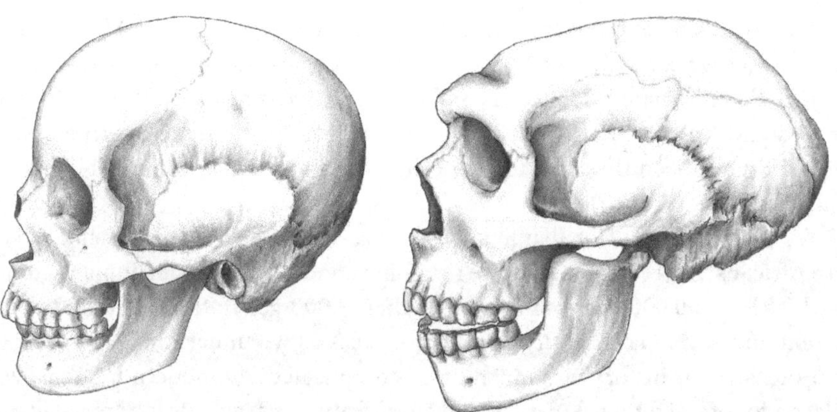

**Figure 10.3** Human and Neanderthal skull morphology. Katherine Wonder (own work).

Hybridization is an important part of our history and has contributed to much of the variation we see within and among human populations today.

If we compare the skulls of humans and Neanderthals, we can see that they are very similar in most ways (Figure 10.3). In fact, when the first Neanderthal skeleton was found, it was thought to be a crime scene until a paleontologist examined the bones and identified a number of characteristics that differ from most humans. Neanderthals have a knob on the back of their skull (*occipital bun*), and they have a prominent brow ridge. It's also true that many humans have jutting brows similar to Neanderthals—including Charles Darwin. But there are other Neanderthal characteristics that differ from most humans, including a weak chin, a lower cranial dome, and an elongated skull. If you look at facial reconstructions of Neanderthals based on their skull morphologies, however, you might notice that they look a lot like humans. While the average of these characteristics differs between humans and Neanderthals, there is wide variation for these traits among humans living today that overlaps with the characteristics of Neanderthals, which probably reflects our history of interbreeding.

Our ancestors who migrated out of Africa around 70,000 years ago looked very different from the Neanderthals they encountered. For one, Neanderthals had light-colored skin and long flowing hair, and the men had full beards. They were stouter and had thick arms and legs, while the humans were smaller boned and less muscular. Our ancestors were mostly hairless with coal-black skin, thin beards, and Afro-natural hair like modern Africans. Imagine what it would have been like to meet people of a different species. Humans migrating out of

Africa would have encountered Neanderthals once they reached the Middle East. From what we can tell, there weren't major battles between these two species, though there is some evidence of conflicts. But there is more evidence that they traded and learned skills from each other. It is clear that they interbred, and it would have been the same for the Denisovans and other hominins living in Eurasia.

Yes, that means that for all humans, our ancestors looked very much like modern Africans. But remember that we're talking about the history of humans going back at least 60,000 years—that's more than 3,000 generations. You and your recent ancestors may identify as different races as we understand them today, but our ancient history was much more complicated. As modern humans, we place a lot of weight on skin color and facial features as indicators of racial identity, but these characteristics can easily respond to selection, inbreeding, and random changes, and they have changed a lot over our history. Let's look at skin color as an example.

Worldwide, there is a general trend for indigenous people living near the equator to have high melanin content in their skin to protect them from the intense sunlight and harmful effects of UV radiation. It's reasonable to assume that our ancestors from eastern Africa also had black skin. As early hominins migrated north, there would have been strong selection to reduce the concentration of melanin in their skin because they would have been wearing more clothing and staying in shelters to keep warm. Our skin needs sun exposure to produce vitamin D. Vitamin D deficiency causes rickets, which has severe symptoms including skeletal deformities and muscle weakness. Consequently, there was strong selection to reduce skin melanin content in northern latitudes because high melanin content would have reduced the ability to produce vitamin D from exposure to sunlight, which made them more susceptible to rickets.[6] Vitamin D production would have been reduced further as they covered more of their skin to keep warm. In recent history there are other examples of high frequencies of rickets due to lack of sunlight exposure. At the beginning of the twentieth century, the Industrial Revolution caused overcrowding and poor air quality, which led to a high frequency of rickets.[7] In regions near the equator, however, people have no problem producing plenty of vitamin D, but exposure to intense sunlight and ultraviolet radiation results in selection for high concentrations of melanin, so people living in equatorial regions have black skin.

---

[6] Z. Hochberg and I. Hochberg, "Evolutionary Perspective in Rickets and Vitamin D," *Frontiers in Endocrinology (Lausanne)* 10 (2019): 306.

[7] B. J. Wheeler et al., "A Brief History of Nutritional Rickets," *Frontiers in Endocrinology (Lausanne)* 10 (2019): 795.

We know from their DNA that Neanderthals had pale skin. They sometimes had blue eyes and often had light-colored hair, including blonde and red hair. In contrast, early humans would have looked very much like modern Africans. Their skin was very dark, and their hair was black and Afro-natural instead of straight or wavy. This makes sense because head hair that becomes long and covers the forehead, neck, and back inhibits the effectiveness of evaporative cooling. Afro-natural hair does not drape down the neck and back and breaks easily so it does not get too long—we can interpret this as an adaptation to improve the effectiveness of evaporative cooling. Male Neanderthals have much more body hair and large bushy beards, which is consistent with the idea that body hair is an adaptation for keeping warm in colder climates. We can be pretty certain that our early human ancestors looked like modern Africans because they lived close to the equator, and people native to eastern Africa today are directly descended from them. Remember that modern Africans from the central and southern regions of the continent do not carry remnants of Neanderthal or Denisovan DNA, and people from other parts of the world do.

The encounter between our ancestors and their Neanderthal relatives had fortunate effects as humans would have learned new skills necessary for survival in colder climates. Interbreeding led to the rapid spread of genes for lighter skin pigmentation in human populations, which increased their ability to produce vitamin D to avoid the debilitating consequences of rickets. The legacy of Neanderthal ancestry in people in Europe and Asia is their light-colored skin and hair, in contrast to the black skin and afro-natural hair of their ancestors who migrated out of Africa some 70,000 years ago.

Does that mean that people who have light-colored skin are part Neanderthal? In a sense, yes—most people from Europe, Asia, and North and South America have some fraction of Neanderthal and Denisovan DNA in their genomes. However, even if the ancestors of people from these regions had not interbred with other hominins, there still would have been strong selection to reduce the melanin content of their skin. After all, the ancestors of Neanderthals migrated out of Africa, so they originally had black skin as well. It's not just about who our ancestors interbred with; it's also about where they lived.

Does that mean that part of the reason for the differences among people originating from different regions of the world is that our ancestors hybridized with different hominin species? We don't know for sure for all of the differences among people, but for Europeans, we know that pale skin, freckles, light-colored hair, large bushy beards, and generally much more body hair were inherited from Neanderthals. On the other hand, having these characteristics does not reflect what proportion of Neanderthal DNA you have. It's possible that characteristics of people originating from Asia, Oceania, and America were inherited from interbreeding with Denisovans or other hominins. But remember, we said that humans were very inbred and there may have been selection for different characteristics in different geographic regions. Inbreeding over many generations enhances the effects of random genetic differences, so each clan could become distinguished by different characteristics. As individual clans colonized different regions, their descendants would have inherited the same features. Differences among people from different places in the world may be partly due to historical interbreeding with archaic hominins. Selection could have played a role as well, and we cannot rule out the possibility that many of these differences could be due to random shuffling of genes among clans colonizing different geographic regions.

We know that there was a lot of hybridization going on thousands of generations ago. But we still haven't solved one final puzzle: What happened to all of those other hominins? Why aren't they still around today? It's possible that their populations declined because they were susceptible to diseases that our African ancestors introduced. For example, we know that contact between European colonists and Aboriginal people from North America and other continents and islands around the world often resulted in high levels of mortality due to unintentionally introduced diseases.[8] It's also possible that our ancestors competed with them for scarce resources. As we discussed previously, we know that the spread of humans out of Africa was relatively rapid, suggesting that they had fast population growth rates, so other hominins may have just been overwhelmed by our ancestors. But the causes of the extinction of Neanderthals, Denisovans, and other archaic hominin species remain a mystery; the only thing we know for sure is that today we carry their genetic legacy in our own genomes.

What about hobbits? We mentioned previously the discovery of a small hominin known as the Flores hominin (*Homo floresiensis*) on the island of Flores in Indonesia. The first thing you need to know is that they weren't really hobbits, so sorry for the disappointment. The fossils of *H. floresiensis* tell us that these hominins were only about three and a half feet tall, but their small size was not due to any genetic disorder. At first the fossils were thought to be only about 12,000 years old, but more recent dating has pushed that back to 50,000 years ago,[9] so it appears that they disappeared soon after the arrival of modern humans on their island during their coastal migration through Indonesia and eventually into Australia. We don't have DNA from their fossils, but based on morphology,

---

[8] H. Pringle, "How Europeans Brought Sickness to the New World," *Science News*, June 4, 2015.
[9] T. Sutikna et al., "Revised Stratigraphy and Chronology for *Homo floresiensis* at Liang Bua in Indonesia," *Nature* 532 (2016): 366–69.

the Flores hominin appears more similar to *H. erectus* than to the more modern lineages of hominin.[10]

But why were they so small? It's actually common to see smaller mammal body sizes on islands, including hippopotamuses and mammoths. And it's interesting to note that a human Pygmy population currently inhabits the island of Flores, but there is not any genomic evidence that they interbred with *H. floresiensis*, so they independently evolved their small stature on the same island.[11] Evolution of small stature is common on islands simply because there is less food available, so the smallest individuals have the highest survival. On the Channel Islands off the coast of California, you can find fossils of pygmy mammoths that were much smaller than the mammoths on the mainland. The size of the Flores archaic hominin is consistent with this trend.

Our final takeaway message is that we can trace the history of all humans around the world to a common origin in Africa. The differences we see among people from different regions of the world represent different histories of migration and hybridization with other human species, as well as inbreeding and random shuffling of genes among clans. But these differences are really very small; in fact, differences among domestic dog breeds is 7.4 times greater than differences among all human populations around the world.[12] In other words, humans of different races are much more closely related to each other than purebred dogs of different breeds. Differences among people whose ancestors originated from different parts of the world probably appear more prominent to us because selection on our ancestors favored their ability to distinguish among individuals based on facial features. There may have been an advantage to recognizing other individuals as social interactions and cooperation became more important for the survival of our ancestral clans. The truth is that the level of genetic differences among populations of humans from different regions is very small compared to variations within other species.[13]

*Summary*—We started out this chapter by reviewing the migrations of different hominin species including early migrations by *H. erectus* and *H. heidelbergensis* and later migrations of *H. sapiens* as they spread out of Africa around 70,000 years ago to colonize all continents around the world except Antarctica. We focused on the interactions between our ancestors and other hominin

---

[10] L. R. Berger et al., "Small-Bodied Humans from Palau, Micronesia," *PLOS One* 3 (2008): e1780.

[11] S. Tucci et al., "Evolutionary History and Adaptation of a Human Pygmy Population of Flores Island, Indonesia," *Science* 361 (2018): 511–16.

[12] B. M. Donovan et al., "Humane Genomics Education Can Reduce Racism," *Science* 383 (2024): 818–22.

[13] A. H. Goodman, *Race: Are We So Different?* (Wiley-Blackwell, 2020).

species including exchanges of knowledge and interbreeding, which resulted in the transfer of their genetic material into our ancestral lineages. We discussed how hybridization has been advantageous for many species of plants and animals and probably facilitated the adaptation of our ancestors to colder, more northern habitats. It's difficult to know for sure what happened to the other species of hominin—why they went extinct—but it's likely that they were simply overwhelmed by the number of humans invading their territories. It's unlikely that they were wiped out by battles or competition; rather, they were probably simply assimilated into our human lineage. The legacy of encounters between our ancestors and other archaic hominins is evident in our DNA and in the characteristics of humans indigenous to Eurasia, North Africa, Australia, and North and South America.

We ended the chapter by briefly discussing the diminutive Flores hominin, which died out when anatomically modern humans invaded the region around 50,000 years ago. This species was only about three and a half feet tall and a descendent of *H. erectus*. Their small size was probably an adaptation to limited food availability on the island where they lived.

# 11
# Settling in

*Many seasons passed, and for the first time in her life, Launua's clan had remained in the same valley for a long time. They had no reason to move because there was plenty of fish to catch from the large lake and from the river that ran through the valley. There was also good food from the land, including grain from one of the grasses and roots from another plant. Launua learned that if she scattered some of the grain on the bare soil, more grass would grow and make more grain. When she started to dig out other plants to make more room for the grass, her fellow clan members were confused. They didn't understand why Launua was throwing good grain onto the ground. But when they saw all the grass that grew for the next harvest, they helped her dig out more plants, and Launua showed them how to scatter the grain on the bare soil to make more grass. The clan now had a steady source of food and their numbers grew.*

*Looking Down the Tree.* Mitchell B. Cruzan, Oxford University Press. © Oxford University Press (2025).
DOI: 10.1093/9780197805190.003.0011

*There were other changes. One was that many other clans had moved into nearby valleys. A few clans of the bone-skinned people remained, but many more people like Launua's clan had migrated into the area. There was trading up and down the large lake's shore, so they had access to a variety of tools. Ornaments, like the string of flat stones, were becoming more common. The clan also looked different. Some said it was because the children played in the mud along the river, so their skin became the same color. But Launua remembered the bone-skinned woman who had joined the clan. She died in childbirth, but her son had the mud-colored skin and he had many children who shared his skin color. Now Launua was one of the only clan members with skin the color of charred wood.*

*Launua was now a clan elder. She had seen more seasons than all the other clan members and spent more of her time showing others different skills. She was often consulted to help make decisions for the clan and resolve disputes. She had lived a long time—longer than most—but she thought it must just have been good luck. In the past she had had a few scrapes with death. One time the river flooded and she almost drowned, and another time an animal attacked and killed one member of the group while they gathered grain. She knew she was fortunate to have lived so long, and she was grateful to be able to see her clan grow strong after all the many seasons when they had struggled just to survive.*

We have talked a lot about the increase in brain size, but we never discussed why humans have larger brains. It's hard to know exactly how our brains differ from earlier hominins. We can measure the total brain volume in fossil skulls, but that does not necessarily provide us with much information about the sizes of different parts of the brain. We do know that Neanderthals generally had larger brains than we do, and based on comparisons of the skulls of their children, it appears that the development of their brains was more similar to chimpanzees.[1] We know that brain size and the number of neurons present are not necessarily indicative of cognitive ability. For example, we have about 100 billion neurons, compared to an estimated 85 billion in Neanderthals and over 250 billion in elephants. It's not about overall size but rather what parts of the brain are larger.

---

[1] P. Gunz et al., "Brain Development after Birth Differs between Neanderthals and Modern Humans," *Current Biology* 20 (2010): R921–R922.

In the case of modern humans, even though our ancestors of 300,000 years ago looked very similar to us, their brains were quite different, and we do not see modern brains developing until sometime within the last 100,000 years.[2]

It's possible that our brains have been shrinking since the Pleistocene. One study suggested that we have smaller brains compared to our ancestors who lived just 10,000 to 20,000 years ago, but this idea has been disputed.[3] A variety of speculative reasons have been proposed for the reduction in brain size, including self-domestication and less need for memory as we developed mechanisms to store information as written records.[4] It's more likely that the reduction in skull size is due to the warming climate and less need for heat retention—similar to the reason that Neanderthals, who lived in cold climates during the Pleistocene glaciations, had larger brains, while *Homo sapiens* from warmer climates did not. But this idea is also speculative.

Comparison of the differences between our brains and those of our ancestors suggests that we have increased language and cognitive skills. We know that large brain size in early species of *Homo* was made possible because of a change in diet; it was enabled because our ancestors started cooking their food over fires. This was important because larger brains require greater amounts of food energy, and cooking food allows our digestive tracts to extract much more energy than we can get from raw food. As we discussed previously, pelvic girdle size is a constraint on brain volume, and infants have rapid brain growth after birth. This constraint caused our infants to be born at an earlier developmental stage, which had a cascade of effects on human characteristics and behaviors.

One of the most important characteristics that contributed to the uniqueness of our ancestors leading up to modern humans was the increasing ability to transfer skills and knowledge among individuals and from one generation to the next—a process known as *cultural transmission*. Older individuals could teach the young skills for tool making as well as hunting and gathering, and adults could transfer information and skills to other adults. Individual skills can be considered as traits that undergo *cultural evolution* as they are passed from one individual to the next.[5] Compared to chimpanzees, humans have an uncanny capacity to mimic and to learn new motor skills through observing the actions of others.[6] This ability was critically important for cultural evolution in our ancestors and

[2] S. Neubauer et al., "The Evolution of Modern Human Brain Shape," *Science Advances* 4 (2018): 5961.

[3] B. Villmoore and M. Grabowski, "Did the Transition to Complex Societies in the Holocene Drive a Reduction in Brain Size? A Reassessment of the DeSilva et al. (2021) Hypothesis," *Frontiers in Ecology and Evolution* (2021): 10.3389.

[4] D. H. Bailey and D. C. Geary, "Hominid Brain Evolution," *Human Nature* 20 (2009): 67–79.

[5] A. Mesoudi et al., "Perspective: Is Human Cultural Evolution Darwinian? Evidence Reviewed from the Perspective of the Origin of Species," *Evolution* 58 (2004): 1–11.

[6] F. Subiaul, "What's Special about Human Imitation? A Comparison with Enculturated Apes," *Behavioral Sciences* 6 (2016): 13.

continues to be critical for the development of technologies in modern societies. The enhanced ability of humans to mimic became apparent in the 1930s when two scientists, Luella and Winthrop Kellogg, raised an infant chimpanzee along with their infant son.[7] They expected Gua, the chimpanzee, to learn from her brother and to become more humanlike over time. Instead, their son, Donald, proved to be better at mimicking and developed vocalizations and behaviors—including knuckle-walking—that he learned from his sister. The experiment was ended because the Kelloggs feared for their son's safety, and Gua died a year later from pneumonia after being returned to a primate center.

Unlike genetically determined traits that must be inherited from one generation to the next, cultural traits can spread much more rapidly among individuals both between generations (*vertical transmission*) and among peers (*horizontal transmission*). Cultural evolution occurs as skills are modified and improved through small variations introduced by individuals, so it is not limited by random changes like genetic mutations; the process of cultural evolution is directed toward continual improvement. Consequently, cultural evolution can lead to rapid behavioral adaptation and was probably intrinsic to the success of our ancestors.

It's probable that there was strong selection on our early ancestors to improve their cognitive skills, and consequently the efficiency and effectiveness of cultural evolution. This process was likely initiated by the fortuitous change in the morphology of early hominin hands that came about as a result of correlated responses to selection on feet to improve bipedal locomotion in the savanna. Bipedalism freed up the hands of our ancestors to engage in activities other than locomotion and generated a hand morphology that was much more adept at handling and manipulating objects. This improved the physical ability to fashion wood, bone, and stone into tools and weapons was coupled with cultural transmission of skills to improve the manufacture of these objects. Hence, there was interaction and feedback between genetic traits and cultural evolution as selection for improved tool-making skills affected the evolution of hand morphology for the improved physical ability to efficiently and effectively manipulate objects. A crucial aspect of this process was the ability of individuals to act as both mimickers—learners—and teachers, and this likely led to selection for improved cognitive ability and, consequently, larger brains.

The process of more rapid cultural and genetic evolution in our ancestors probably did not begin until after they began to use fire to cook their food, which enabled the evolution of larger brains. Our *Australopithecus afarensis*-like ancestors had hand and wrist morphologies that allowed them to fashion

---

[7] W. N. Kellogg and L. A. Kellogg, *The Ape and the Child: A Comparative Study of the Environmental Influence upon Early Behavior* (Hafner Publishing Co., 1933).

simple tools similar to chimpanzees, but their capacity for cultural evolution may have been limited by their cognitive ability. There is no evidence that they had controlled use of fire, and their brains were only slightly larger than those of modern chimpanzees. It was not until sometime later—around two million years ago and early in the origin of the *Homo* species—that there is evidence for controlled use of fire,[8] which in turn led to the potential for the evolution of much larger brains. It was most likely selection for improved cultural transmission of skills and knowledge that resulted in selection for increased cognitive ability and brain size.

Our *Homo erectus* ancestors' brains were about twice the size of Lucy's, but they were still one-third smaller than those of *H. sapiens*. Apparently, the much larger brain of *H. erectus* was associated with improved cultural transmission, which enabled their spread across Africa and Eurasia. Improvements in individual clans' skills for making tools and weapons would have led to increased food supplies and, consequently, increased population growth such that the more highly skilled clans would have spread more quickly and displaced others.

Hence, it appears that the origin and evolution of the lineage leading up to modern humans were the result of a fortuitous combination of traits that promoted improved tool-making skills and cultural evolution, and consequently, improved cognitive abilities enabled by much larger brains. An important aspect of this process was improved oral communication abilities. As our ancestors improved their language skills, their ability to pass on knowledge to the next generation also increased. The development of advanced tool-making skills required more precise cultural transmission and was apparently associated with the origin of spoken language around 1.75 million years ago.[9] Eventually, the transmission of knowledge became even more efficient when our ancestors developed the skills to record knowledge in a form that was more permanent and could be interpreted by others—written language. The first evidence of a written language was only around 5,500 years ago, so prior to that, the transmission of skills and knowledge primarily occurred through mimicking and oral communication. Verbal traditions and storytelling were the primary mechanisms for the transmission of knowledge over much of the history of our ancestors, and it is only in the last few hundred years that literacy has become widespread and common for the majority of people in cultures around the world.

[8] S. R. James, "Hominid Use of Fire in the Lower and Middle Pleistocene: A Review of the Evidence," *Current Anthropology* 30 (1989): 1–26.

[9] N. T. Uomini and G. F. Meyer, "Shared Brain Lateralization Patterns in Language and Acheulean Stone Tool Production: A Functional Transcranial Doppler Ultrasound Study," *PLOS One* 8 (2013): e72693.

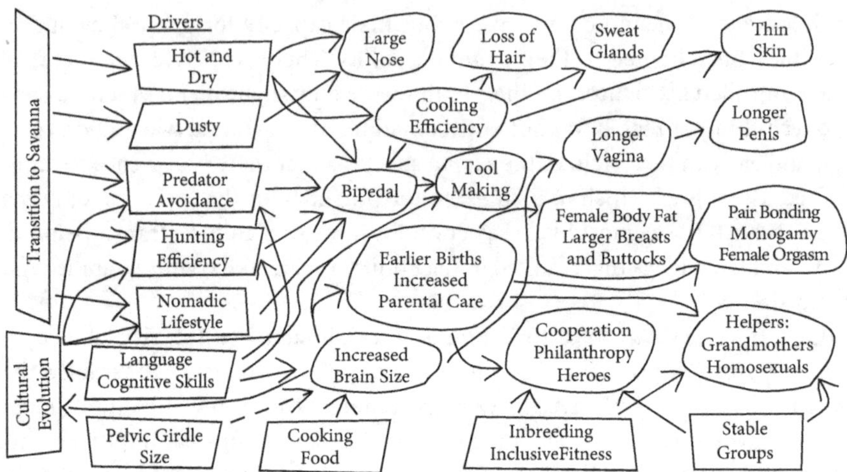


**Figure 11.1** Concept diagram for the drivers of selection (in boxes) and resulting unique human characteristics (in ovals). M.B. Cruzan (own work).

In previous chapters we discussed a number of characteristics that are unique to humans and some that are shared with our closest living relatives. You may have noticed that there are a lot of connections from one characteristic to another, so the evolution of the modern human form is the result of a number of things working together. One way to visualize the relationships among different characteristics is to develop a concept diagram that summarizes the evolution of human traits (Figure 11.1). In this diagram, you'll notice that the labeled boxes contain the drivers—the primary selective forces—and arrows point to the characteristics that were affected by them.

Let's start by thinking about what happened when dry forests and savannas replaced much of the wet forests in eastern Africa starting as early as ten to fifteen million years ago. The earliest australopithecine populations colonizing savanna habitats would have been under strong selection for adaptation to locomotion on flat ground and surviving under hot, dusty, and dry conditions. Along the left side there is a box labeled "Transition to Savanna" with arrows pointing to a box labeled "Hot and Dry" as well as other boxes labeled "Dusty," "Predator Avoidance," "Nomadic Lifestyle," and "Hunting Efficiency." These are the drivers of selection, and we can see how they affected the evolution of different human traits (in ovals). For one, there are arrows from the "Hot and Dry" and "Dusty" boxes to an oval containing "Large Nose." Next, notice that an arrow from "Hot and Dry" points to "Cooling Efficiency," which in turn has arrows pointing to ovals for the loss of hair, the increase in the number of sweat glands, and the evolution of thin skin underlain by blood vessels, all of which improved cooling efficiency.

The next thing you might think of is that our australopithecine ancestors became bipedal during their transition to savanna habitats. To account for this, there are arrows pointing from several of the driver boxes to an oval labeled "Bipedal," which in turn initiated the evolution of grasping hands and improved our ancestors' tool-making abilities. Selection for improved tool-making skills, cultural evolution, and consequently larger brains would have been more likely to be operative in the descendants of our australopithecine ancestors in the genus *Homo*. Note that there is also an arrow from the "Increased Brain Size" oval to the "Tool Making" oval, indicating that improved cognitive skills were required for advanced tool making.

An increase in brain size in our *Homo* ancestors meant that their infants required more parental care. This resulted in the evolution of traits and behaviors that improved offspring survival such as monogamy; nonreproductive helpers, which included homosexuals and grandmothers, and cooperation within the group. The evolution of cooperative behaviors required two things. One is inbreeding, which increases the advantages of cooperation through inclusive fitness. The other is something that we have not discussed yet but is also critical: *stable group composition*. This is important because within stable groups, individuals know each other and understand that cooperation is rewarded because favors are returned in the future.

Some of the changes in our ancestors' characteristics came about more because of evolutionary constraints and coordinated evolution and did not necessarily improve survival. For example, as head size increased, the uterus had to be higher up in the abdomen, and because of that, females of our species have longer vaginas, and ultimately males have longer penises as well. But larger head size resulted in earlier births so that our infants required many years of care instead of just a few months, as in Lucy's species. This led to increased fat deposits including enlarged breasts and buttocks in women, monogamy, elaboration of the clitoris to improve functionality for female orgasm and to promote pair bonding, and traits that favored cooperation, including the inclusion of helpers such as grandmothers and homosexual individuals in stable groups. Much of our social and sexual behavior is a product of the fact that the infants of our ancestors required care for many years, and because they lived in clans of inbred individuals, cooperation was favored through inclusive fitness.

The concept diagram (Figure 11.1) provides a summary of how the environment selects for adaptive traits and behaviors and makes it easier to see how selection on one trait can have cascading effects on others. The single most influential event in the history of our lineage was the transition to savannas and selection for bipedalism, which caused correlated changes in the hands of our ancestors.[10] This event—which freed up the hands so they were no longer needed for locomotion and made them more adept at grasping—opened the door to selection for a more refined ability to manipulate objects and eventually to fashion wood, stone, and bone into tools. Once our ancestors had opposable thumbs, those clans that were more skilled at tool making were more successful, and selection continued to refine hand morphology to improve handling of objects through both cultural and genetic evolution. The initiation of tool making with an improved hand morphology is what most likely set the stage for selection to improve our ancestors' cultural transmission skills, and cultural evolution became increasingly important. Selection for improvement of cultural transmission skills was the primary selective driver for increasing brain volume and cognitive skills in our *Homo* ancestors. This second major event in our history—the dramatic increase in brain size along with inbreeding in clans—had a multitude of consequences, including monogamy and female orgasm, increased fat deposits and larger breasts and buttocks in females, increased cooperation, and the inclusion of homosexual helpers. But it was really selection for bipedalism on the savanna and its fortuitous consequences for hand morphology that initiated the whole sequence of events leading up to the appearance of anatomically modern humans with their superior cognitive skills.

It's important to recognize that all of the advances that were made as natural selection worked on our ancestors to eventually produce modern humans came at a tremendous cost. Evolution by natural selection occurs via small improvements from beneficial genetic mutations over long periods of time. Each time a beneficial mutation appeared in populations of our ancestors, it was carried by a single individual and had to increase in frequency and spread until all individuals of the species carried it. Individuals carrying the new mutation had higher chances of survival and produced more children, while those without the beneficial mutation had lower chances of survival. In other words, many individuals died as the new mutation increased in frequency and each improvement was made. When we consider the numerous beneficial mutations that contributed to the generation of our species, we realize that many millions—probably billions—of individuals had to die as small improvements accumulated generation after generation over the past four to six million years. Perhaps we should not look down on our ancestors as savage and brutish individuals that were "less than human" but rather view them with some reverence, respect, and thankfulness for their sacrifices and resilience in the face of tremendous challenges.

[10] C. Rolian, "Re-evaluating the Correlated Evolution of Human Hands and Feet Using Viability Selection Modeling," *American Journal of Biological Anthropology* 183 (2024): e24693.

At this point you might be asking whether our species is still subjected to natural selection: Are we continuing to evolve? Will our future species have even larger heads, as is portrayed in some science fiction books and movies? Remember that for natural selection to cause changes, weaker individuals have to either die or never reproduce as the stronger proliferate. If anything, the opposite is true today; in modern societies we have done everything we can to ensure the survival of every individual. In many societies around the world, we are buffered from selection by controlled environments and advanced medical technologies. Consequently, there are many individuals alive today who would have never survived beyond a young age just a century ago. While all of us living in technologically advanced societies are grateful to be able to enjoy much longer lives, there are also some negative consequences of our medical advancements. By buffering ourselves from natural selection, we have allowed many harmful mutations to increase and spread, and some evolutionary biologists have warned that we can expect declines in health and longevity in future generations.[11] A decline in fitness is inevitable in human populations with access to modern medicine unless we can develop gene editing technologies that become inexpensive and widespread. Even though technologies such as CRISPR-Cas9 are already being applied to correct genetic deficiencies, it's important to realize that these treatments only affect the living individual; their children will inherit the same faulty genes that afflicted them. This is true because current technologies only affect the genomes of cells in the body other than the gonads. Consequently, gene editing will continue to be necessary to improve the health of every future generation unless methods can be developed to repair genomes of the human germline,[12] which is the source of sperm and eggs. Meanwhile, deleterious mutations will continue to accumulate in populations with access to modern medicine, and tackling these problems will be a burden that future generations will inherit.

Natural selection on our ancestors not only affected their appearance but also had consequences for the health and longevity of modern humans. Recall that throughout our ancestral history individuals had short lifespans; it was only in the past few hundred years that people began living past the age of forty. Consequently, our "young bodies" have been fine-tuned by selection over the past several million years, but selection has not improved our health in later years. As we age, we are subject to health declines and maladies such as heart disease, cancer, diabetes, hypertension, osteoporosis, Alzheimer's disease,

[11] M. Lynch, "Rate, Molecular Spectrum, and Consequences of Human Mutation," *Proceedings of the National Academy of Sciences* 107 (2010): 961–68.

[12] National Academies of Sciences, Engineering, and Medicine et al., *Human Genome Editing: Science, Ethics, and Governance* (National Academies Press, 2017).

and many more. Much effort has been put into research and medical developments to increase longevity. For example, we know that the caps on the ends of chromosomes—*telomeres*—gradually erode as we age. Many millions of dollars have been invested in developing treatments to maintain telomere length, but it turns out that telomere shortening is more likely to be an effect of aging rather than a cause.[13] Declines in health and the higher instances of disease later in life are more likely consequences of the short lifespans of our ancestors; without the influence of selection, mutations having negative effects past the age of forty were free to accumulate.[14] While we have benefited in many ways from the consequences of selection due to the challenges our ancestors faced, we also carry the legacy of the absence of selection later in life. Unfortunately, we all face a myriad of diseases and decline that will afflict us as we age. There is no "silver bullet" to increase longevity; instead, senescence in humans is more like "death by a thousand cuts."

Inbreeding and the evolution of cooperation within clans were critical for the survival of our ancestors, but they may have led to some of the darker aspects of our behaviors and history. As members of a clan were loyal to each other, they may have become fearful and distrustful of individuals from other clans. Clans may have become more territorial and more likely to have conflicts with others in environments where resources were limited. We know that there was trading and transfer of technologies between our human ancestors and Neanderthals, but there is also evidence of violence. Skulls from these regions of overlap sometimes show evidence of damage from blunt weapons such as clubs, and individuals have been found with broken forearms, suggesting they were defending themselves from attack.

While conflicts among individuals may have occurred regularly, there are different opinions on whether organized war between groups is just a human characteristic or a symptom of more complex societies that have developed only in recent history.[15] It's true that evidence of large-scale conflicts is only available from the last few thousand years, but this does not preclude the possibility that war between groups competing for resources may have occurred throughout the history of our species. The tendency for organized conflicts in recent history was exacerbated by the accumulation of resources in established settlements, which then became tempting targets for competing groups to acquire by force. Even though clan loyalty resulted in cooperation, philanthropy, and other generous

---

[13] A. Shekhidem et al., "Telomeres and Longevity: A Cause or an Effect?," *International Journal of Molecular Science* 20, no. 13 (2019): 3233.

[14] J. Rodríguez et al., "Antagonistic Pleiotropy and Mutation Accumulation Influence Human Senescence and Disease," *Nature Ecology and Evolution* 1 (2017): 0055.

[15] R. B. Ferguson, "War Is Not Part of Human Nature. War May Not Be in Our Nature After All," *Scientific American*, September 1, 2018.

human behaviors, we should face the fact that it's probably responsible for the phenomena of tribalism and nationalism, which have led to conflicts, genocide, wars, and the dehumanization of others.

We've covered a lot of ground in this book. We've learned about the history of our lineage going back through earlier species of *Homo* to australopithecines and ultimately to our common ancestors with chimpanzees. By applying the principles of evolutionary biology, we have developed some novel ways of thinking about human evolution, including how (1) selection for adaptation to the savanna in australopithecines led to the origin of bipedalism and grasping hands; (2) selection for more effective cultural transmission led to larger brains and ultimately verbal and written languages; (3) increasing brain size in our earlier *Homo* ancestors resulted in more fat deposits in women, monogamy, and female orgasm for improved pair bonding; and (4) inbreeding in clans favored cooperation and the inclusion of nonreproductive helpers—grandmothers and homosexuals.

By this point you should see that the unusual appearance and behaviors of humans compared to other animals can be explained by the evolutionary processes of natural selection and cultural transmission. The first critical step that created the cascade of selection processes leading up to modern humans was the transition from heavily forested areas to grassland and savanna habitats. Hence, we are a product of the environments our ancestors lived in and the selective pressures they were subjected to. In many ways we are just like other animals, but in our case, selection for increased language and cognitive skills in our ancestors resulted in a species with immense capacity for creation and destruction. We have an unending desire to modify our environments to fit our needs. We create marvelous technologies and structures, and in the process, we damage and pollute the natural world that spawned us.

We now face one of the greatest challenges of our species' existence—the preservation of the ecosystems we rely on. Just as our ancestors had to fight for their survival on a daily basis, we may be facing a similar challenge once again, but this time we are fighting for the survival of our planet as we know it.

Sometimes it feels like we are facing overwhelming challenges, but keep in mind that the challenges we face today really pale in comparison to what our ancestors had to deal with. We have outdone ourselves; our constant motivation to mold our environments to our will has yielded unforeseen consequences at a global level. Our planet may never be the same as it was just a few hundred years ago. But while it may be true that our future world will be different than the one we live in today, those things that have always mattered most—our health and happiness and the health and happiness of our friends and families—will remain the same.

The most important takeaway from our human story is that we are a resilient species that has flourished not only because of our superior language and cognitive abilities but also because of our immense capacity for empathy and cooperation. Our history illustrates that we have connections to each other and to all other species we share this planet with. We are all members of the same human family, and cooperation across the entire human clan will be critical to ensure healthy and comfortable futures for all of us.

*Summary*—We started this chapter by discussing how bipedalism had cascading consequences for our ancestors. First, it freed their hands, which were now much more adept at manipulating objects, so they could begin to create tools from wood, bone, and stone. It must have taken some time, but eventually selection began to favor improved cognitive ability for more effective cultural transmission of tool-making skills. This process was accelerated once our ancestors began cooking their food, which enabled the evolution of much larger brains. Cultural evolution had consequences for genetic traits, and especially brain volume, as those clans with more effective teaching and learning abilities grew and spread as they repeatedly split to form new clans. Early on, cultural transmission occurred primarily through mimicking, but improved cognitive skills would have allowed deliberate teaching and improved learning. Eventually, vocalizations evolved into spoken languages, which became more proficient so knowledge could be transmitted through storytelling. It was only within the last few thousand years that written languages emerged, which allowed for the recording of knowledge in a permanent form.

We pulled together the information we previously discussed in a concept diagram to emphasize the interactions among historical environments, traits, and behaviors and the consequences for the evolution of our ancestors. First, we reviewed the effects of the transition to savanna habitats on our australopithecine ancestors. Selection for bipedalism had consequences for their hands, which fortuitously provided them with a superior ability to manipulate objects and perhaps to fashion simple tools. Their new habitat had consequences for their facial appearance as the dry and dusty conditions selected for increased nasal areas and the prominent nose that we are familiar with today. Living on the savanna meant that they became mostly furless with sweat glands distributed over most of their bodies and thin skin with high densities of blood vessels for efficient evaporative cooling. Their skull size was not much larger than their chimpanzee cousins, so infants were breastfed for only a few months and females did not have large fat deposits in their breasts and buttocks like their larger-skulled descendants did. They probably lived in polygynandrous groups of multiple males and females.

Increasing brain volume during the transition from an australopithecine ancestor to *Homo* constrained fetal development, so infants became increasingly altricial. Increased care was required to ensure the survival of offspring, which had multiple consequences for the behavior and appearance of our ancestors. The time required for breastfeeding was extended from a few months in australopithecines to several years in *H. erectus*. Consequently, selection favored increased fat deposits in females, particularly in their breasts and buttocks. Increasing head size also resulted in the displacement of the uterus upward in the body cavity. As vagina length increased, there was strong selection for increased penis length to ensure the placement of sperm close to the cervix. The necessity for increased infant care favored monogamy and more participation from both partners, and the elaboration of the clitoris for female orgasm improved pair bonding, resulting in stable monogamous relationships. Inbreeding in clans generally favored cooperation and self-sacrifice, which benefited individuals through inclusive fitness because of gene sharing among clan members. The success of clans was improved through the contributions of nonreproductive helpers including exclusively homosexual individuals and perhaps the occasional female who lived beyond her menstrual years. Increased cooperation and the inclusion of homosexuals in clans may have been crucially important for the survival of our ancestors during the severe drought that precipitated a 100,000-year population bottleneck—a near-extinction event—some 900,000 years ago.

It's probable that males and females of our *Homo* ancestors participated in all aspects of clan life including childcare, gathering of grains and roots, food preparation, and hunting and killing animals for their meat and pelts. Since death rates were high, there would have been orphaned children that were adopted by monogamous pairs, including lesbian and gay couples. Unfortunately, it's likely that exclusively homosexual women sometimes would have been victims of rape and would have had children of their own. As we emphasized before, in resource-limited environments, the clans with high levels of cooperation and sharing would have been the most successful and the most likely to survive to leave descendant generations that eventually evolved into *H. sapiens*. It's encouraging that many cultures around the world are beginning to reject doctrines promoting patriarchal societies and discrimination against individuals based on race and gender and sexual identity. Perhaps we are beginning to return to the more natural state of the cultural norms of our ancestors, which was crucial for their survival over the last million years.

Our improved cognitive skills for the transmission and development of knowledge have had enormous consequences for our lives and the lives of all species. We've created marvelous technologies that can be used to ensure long and easygoing lives or to elicit destruction, death, and genocide. We now carry an enormous responsibility for the livelihood of all people and of all living species around the world. Our impact on the atmosphere is precipitating higher temperatures, more frequent and intense hurricanes and typhoons, and more severe episodes of heat waves, drought, and flooding. As some regions become inhospitable due to higher temperatures and rising sea levels, whole societies will be displaced, leading to mass migrations. Whether we as a species can survive the resulting calamity will depend on our ability to see humanity in all of our neighbors and whether we can reinvigorate our core instincts for cooperation, which were critical for the survival of our ancestors.

# Epilogue

*Launua had lived many seasons, and now she was spending much of her time teaching the young members of the clan all the skills she had learned. On this particular day, she was far from camp with a group of youngsters. She had injured her foot the day before so she had some difficulty walking, but she wanted to show them how to find the best places to dig for roots. As she was helping one of them unearth a root, she caught sight of a flash of fur in the corner of her eye. She quickly spun around and managed to hit the animal before it could pounce on one of the children. There was screaming, and the children began to run as Launua faced the animal, swinging her digging stick to fend it off. Once she was sure the children were safe, she turned to run. As she fell to the ground, she saw the horror on the faces of the children as they looked back, but she felt a deep sense of peace knowing that they were safe and that her clan was strong. And then her life ended.*

# Author Notes

I wrote this book as an exercise in science communication. I wanted to pull together the information I have been accumulating over the past twenty years and regularly presenting in my evolution courses, but I wanted to do it in a way that might reach a wider audience than the typical popular science book. My hope was to provide an entertaining context while allowing readers to come to a better understanding of evolutionary biology and how scientists develop ideas based on data and logical arguments, and to provide them with some insights into the origins of our unique human behaviors and characteristics. My goal is to provide clear explanations of evolutionary concepts with minimal jargon. Consequently, the explanations are accurate but not as in-depth or concise as students and scientists who are familiar with concepts in evolutionary biology might have preferred.

I never intended to write fiction. My thought was to describe conditions that existed 70,000 years ago, but this came across as dictatorial and uninteresting. I ultimately decided that the best way to communicate the challenges that our ancestors faced was to create a fictional character who lived around the time of the major migrations of humans out of Africa. The life of *Launua* (pronounced "Lau-nua," from the Esperanto *La Unua*, which translates to "The First One") covers what probably would have been many lifetimes as our human ancestors migrated out of eastern Africa and eventually encountered Neanderthals along the shores of the Red Sea. Her story serves as a backdrop that threads throughout the book to provide a more personal context for the evolutionary concepts discussed. The details of Launua's life and of her clan are based on what we know about the behaviors, technologies, and cognitive abilities of people who lived at that time. Her story is completely fictional; she is not meant to represent any important individual in the history of our species.

The traits and behaviors that I discuss represent some of the most unique aspects of humans. The list is not meant to be comprehensive. The traits discussed in the first few chapters are meant to illustrate how the modern human form was generated from the harsh landscapes that our ancestors inhabited. Other traits and behaviors are included because they fit into the cascade of selective forces and responses that were most critical for the success of our ancestors and for the morphology of modern humans. My goal is to illustrate how multiple selective pressures combined to generate human characteristics and behaviors that are otherwise difficult to understand when we think about them only in the context of the artificial environments that we currently inhabit.

# Index

*For the benefit of digital users, indexed terms that span two pages (e.g., 52–53) may, on occasion, appear on only one of those pages.*